T0134522

Studies in Systems, Decision and Control

Volume 145

Series editor

Janusz Kacprzyk, Polish Academy of Sciences, Warsaw, Poland
e-mail: kacprzyk@ibspan.waw.pl

The series "Studies in Systems, Decision and Control" (SSDC) covers both new developments and advances, as well as the state of the art, in the various areas of broadly perceived systems, decision making and control- quickly, up to date and with a high quality. The intent is to cover the theory, applications, and perspectives on the state of the art and future developments relevant to systems, decision making, control, complex processes and related areas, as embedded in the fields of engineering, computer science, physics, economics, social and life sciences, as well as the paradigms and methodologies behind them. The series contains monographs, textbooks, lecture notes and edited volumes in systems, decision making and control spanning the areas of Cyber-Physical Systems, Autonomous Systems, Sensor Networks, Control Systems, Energy Systems, Automotive Systems, Biological Systems, Vehicular Networking and Connected Vehicles, Aerospace Systems, Automation, Manufacturing, Smart Grids, Nonlinear Systems, Power Systems, Robotics, Social Systems, Economic Systems and other. Of particular value to both the contributors and the readership are the short publication timeframe and the world-wide distribution and exposure which enable both a wide and rapid dissemination of research output.

More information about this series at http://www.springer.com/series/13304

M. Hadi Amini · Kianoosh G. Boroojeni
S. S. Iyengar · Panos M. Pardalos
Frede Blaabjerg · Asad M. Madni
Editors

Sustainable Interdependent Networks

From Theory to Application

 Springer

Editors
M. Hadi Amini
Carnegie Mellon University
Pittsburgh, PA
USA

Panos M. Pardalos
University of Florida
Gainesville, FL
USA

Kianoosh G. Boroojeni
Florida International University
Miami, FL
USA

Frede Blaabjerg
Aalborg University
Aalborg East
Denmark

S. S. Iyengar
Florida International University
Miami, FL
USA

Asad M. Madni
University of California Los Angeles
Los Angeles, CA
USA

ISSN 2198-4182   ISSN 2198-4190   (electronic)
Studies in Systems, Decision and Control
ISBN 978-3-030-08985-6   ISBN 978-3-319-74412-4   (eBook)
https://doi.org/10.1007/978-3-319-74412-4

Printed on acid-free paper

This Springer imprint is published by the registered company Springer International Publishing AG part
of Springer Nature
The registered company address is: Gewerbestrasse 11, 6330 Cham, Switzerland

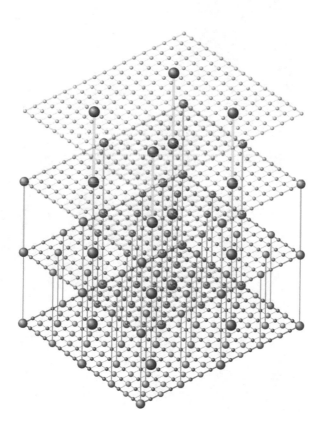

This book is dedicated to President Mark Rosenberg of FIU and Provost Ken Furton for all their dedication to make FIU a great research school for inspiring students of the World.

S. S. Iyengar
Director and Ryder Professor
September 2017

Preface

Since the concept of Internet of Things leading to smart cities, smart grids, and other technologies emerged, there has been a need to adopt to a sustainable way of utilizing the technologies. Researchers have tried to answer this question on the emerging concern regarding the optimal operation of real-world large-scale complex networks. A smart city is a vision of the top brass of researchers to integrate multiple information and communication technologies in a secure fashion to manage a city's assets including transportation systems, power grids, distributed sensor networks, water supply networks, and other community services. Our reliance on these complex networks as global platforms for sustainable cities and societies as well as shortage of global non-renewable energy sources has raised emerging concerns regarding the optimal and secure operation of these large-scale networks. Although the independent optimization of these networks leads to locally optimum operation points, there is an exigent need to move toward obtaining the globally optimum operation point of such networks while satisfying the constraints of each network properly.

This book focuses on the various applications that exist for the many theoretical concepts that exist in the field. In this book, the various concepts related to the theory and application of interdependent networks, optimization methods to deal with the computational complexity of such networks, and their role in smart cities are described.

In the first half of the book, an overview of the interdependent networks and their applications in sustainable interdependent networks is presented. Then, a thorough investigation of both classic and modern optimization/control problems in sustainable interdependent networks is provided. A physical model and corresponding constraints of these networks are also discussed elaborately in the breadth of the book. In the second half of the book, the application of sustainable networks in power microgrids, energy and water networks, and smart cities in specific is highlighted.

To this end, the influential networks including power and energy networks, transportation networks, wireless sensor networks, communication networks, water networks, and societal networks are considered. Our reliance on these networks as global platforms for sustainable cities and societies has led to the need for developing novel means to deal with arising issues. The considerable scale of such networks brings a large variety of computational complexity and optimization challenges.

The book series aims at covering wide areas from theoretical toward practical aspects of interdependent networks. This volume and the upcoming volumes will cover different aspects of interdependent networks categorized in the following five parts:

Part A. Classic Optimization and Control Problems—consisting of research articles from outstanding researchers in the field that discuss and provide an insight into the classical and theoretical problems that exist in optimization and control.

Part B. Efficient Methods for Optimization Problems in Large Scale Complex Networks: From Decentralized Methods Toward Fully Distributed Approaches—A more advanced outlook and possible methodologies that could be incorporated in large networks using various approaches are highlighted through manuscripts from leading researchers in this field.

Part C. Modeling the Interdependency of Power Systems, Communication Networks, Energy Systems (e.g. Gas, Renewables), Transportation Networks, Water Networks, and Societal Networks—The ways in which sustainable interdependence can be achieved across the various networks that span the horizons of smart cities are explained mathematically in terms of models that could be implemented. All the articles in this section deal with the interdependency of at least two networks that are an integral part of realizing a smart city.

Part D. Application of Power and Communication Networks: Co-simulation Platforms for Microgrids and Communication Networks, and Smart Grid Test Beds—The various papers under this topic highlight the state-of-the-art technologies that are in place in the area of smart grids.

Part E. Application of Sustainable Interdependent Networks in Future Urban Development: The Tale of Smart Cities—This section constitutes papers that discuss the aspects to be pondered in the future and the direction in which the current school of researchers are heading toward realizing ideal sustained and interdependent networks in the future.

Pittsburgh, USA   M. Hadi Amini
Miami, USA   Kianoosh G. Boroojeni
Miami, USA   S. S. Iyengar
Gainesville, USA   Panos M. Pardalos
Aalborg East, Denmark   Frede Blaabjerg
Los Angeles, USA   Asad M. Madni
February 2018

Acknowledgements

M. Hadi Amini would like to thank his parents, Alireza and Parvin, for their unconditional support and encouragement throughout his life. He also would like to appreciate the kindly support of his artist sister, Azadeh; his uncle; and his brother-in-law. He also would like to express his gratitude toward his beloved niece, Anoosheh, and his beloved nephew, Arshia.

Kianoosh G. Boroojeni and S. S. Iyengar would like to thank NSF, NIII, and other research funding institutes for funding their research works.

Kianoosh G. Boroojeni would also like to express his sincere gratitude to his family for their continuous inspiration and support.

Panos M. Pardalos acknowledges support from the Paul and Heidi Brown Preeminent Professorship in ISE, University of Florida.

Frede Blaabjerg wishes to thank his team in Aalborg and his family (Ina, Anja, and Jakob) to allow him to use so much time on his technical field, so it has been possible to be a part of this book.

Contents

**1 A Panorama of Future Interdependent Networks: From
Intelligent Infrastructures to Smart Cities** 1
M. Hadi Amini, Kianoosh G. Boroojeni, S. S. Iyengar,
Frede Blaabjerg, Panos M. Pardalos and Asad M. Madni

**Part I Strategic Planning of Developing Sustainable Interdependent
Networks**

**2 Calling for a Next-Generation Sustainability Framework
at MIT** 13
Julie Newman

**3 Toward a Smart City of Interdependent Critical
Infrastructure Networks** 21
Arif I. Sarwat, Aditya Sundararajan, Imtiaz Parvez,
Masood Moghaddami and Amir Moghadasi

**4 Interdependent Interaction of Occupational Burnout
and Organizational Commitments: Case Study of Academic
Institutions Located in Guangxi Province, China** 47
Xiazi Sun

**Part II Solutions to Performance and Security Challenges
of Developing Interdependent Networks**

**5 High Performance and Scalable Graph Computation
on GPUs** 67
Farzad Khorasani

6 Security Challenges of Networked Control Systems 77
Arman Sargolzaei, Alireza Abbaspour, Mohammad Abdullah Al
Faruque, Anas Salah Eddin and Kang Yen

7   Detecting Community Structure in Dynamic Social Networks
     Using the Concept of Leadership   97
     Saeed Haji Seyed Javadi, Pedram Gharani and Shahram Khadivi

Part III   Electric Vehicle: A Game-Changing Technology
           for Future of Interdependent Networks

8   Barriers Towards Widespread Adoption of V2G Technology
     in Smart Grid Environment: From Laboratories to
     Commercialization   121
     Nadia Adnan, Shahrina Md Nordin
     and Othman Mohammed Althawadi

9   Plug-in Electric Vehicle Charging Optimization Using
     Bio-Inspired Computational Intelligence Methods   135
     Imran Rahman and Junita Mohamad-Saleh

Part IV   Promises of Power Grids for Sustainable Interdependent
          Networks

10   Coordinated Management of Residential Loads in Large-Scale
     Systems   151
     Amir Safdarian

11   Estimation of Large-Scale Solar Rooftop PV Potential
     for Smart Grid Integration: A Methodological Review   173
     Dan Assouline, Nahid Mohajeri and Jean-Louis Scartezzini

12   Optimal SVC Allocation in Power Systems for Loss
     Minimization and Voltage Deviation Reduction   221
     M. Hadi Amini, Rupamathi Jaddivada, Bakhtyar Hoseinzadeh,
     Sakshi Mishra and Mostafa Rezaei Mozafar

13   Decentralized Control of DR Using a Multi-agent Method   233
     Soroush Najafi, Saber Talari, Amin Shokri Gazafroudi,
     Miadreza Shafie-khah, Juan Manuel Corchado and João P. S. Catalão

14   Complex Distribution Networks: Case Study Galapagos
     Islands   251
     Diego X. Morales, Yvon Besanger and Ricardo D. Medina

Index   283

Editors and Contributors

About the Editors

M. Hadi Amini is currently a Ph.D. candidate at the Department of Electrical and Computer Engineering, Carnegie Mellon University, Pittsburgh, PA, where he received the M.Sc. in Electrical and Computer Engineering in 2015. Prior to that, he received the B.Sc. from the Sharif University of Technology, Tehran, Iran, in 2011, and the M.Sc. from Tarbiat Modares University, Tehran, in 2013, both in Electrical Engineering. He serves as reviewer for several high-impact journals, international conferences, and symposiums in the field of power systems. He has published more than 50 refereed journal and conference papers in the smart energy systems and transportation network-related areas. He is the recipient of the best paper award of "Journal of Modern Power Systems and Clean Energy" in 2016, best reviewer award of "IEEE Transactions on Smart Grid" from the IEEE Power & Energy Society in 2017, outstanding reviewer award of "IEEE Transactions on Sustainable Energy" in 2017, and the deans honorary award from the president of Sharif University of Technology in 2007. He ranked 26th among 270,000 participants in the Nationwide University Entrance Exam for B.Sc. in 2007. His current research interests include interdependent networks, distributed/decentralized optimization algorithms in energy systems, electric vehicles, and state estimation (homepage: www.hadiamini.com).

Kianoosh G. Boroojeni received Ph.D. in Computer Science from Florida International University in 2017. He received his B. Sc degree from University of Tehran in 2012 and his M.Sc. degree from Florida International University in 2016. His research interests include smart grids and cybersecurity. He is the author/co-author of three books published by MIT Press and Springer. During his Ph.D. years, he has published more than 30 publications in the form of book chapters, peer-reviewed journal papers, and conference papers. He is currently collaborating with Dr. S. S. Iyengar to study some network optimization and security problems in the context of smart urban development.

S. S. Iyengar is a leading researcher in the fields of distributed sensor networks, computational robotics, and oceanographic applications and is perhaps best known for introducing novel data structures and algorithmic techniques for large-scale computations in sensor technologies and image processing applications. He is currently the Director and Ryder Professor at Florida International University's School of Computing and Information Sciences in Miami, FL. He has published more than 500 research papers and has authored or co-authored more than 12 textbooks and edited 10 others. Iyengar is a Member of the European Academy of Sciences, a Fellow of the Institute of Electrical and Electronics Engineers (IEEE), a Fellow of the National Academy of Inventors (NAI), a Fellow of the Association of Computing Machinery (ACM), a Fellow of the American Association for the Advancement of Science (AAAS), and a Fellow of the Society for Design and Process Science (SDPS). He has received the Distinguished Alumnus Award of the Indian Institute of Science.

Panos M. Pardalos is a Distinguished Professor and the Paul and Heidi Brown Preeminent Professor in the Departments of Industrial and Systems Engineering at the University of Florida, and a world-renowned leader in Global Optimization, Energy Systems, Mathematical Modeling, and Data Sciences. He is a Fellow of AAAS, AIMBE, and INFORMS and was awarded the 2013 Constantin Caratheodory Prize of the International Society of Global Optimization. In addition, Dr. Pardalos has been awarded the 2013 EURO Gold Medal prize bestowed by the Association for European Operational Research Societies. He is also a Member of the New York Academy of Sciences, the Lithuanian Academy of Sciences, the Royal Academy of Spain, and the National Academy of Sciences of Ukraine. He has published over 500 papers, edited/authored over 200 books, and organized over 80 conferences. He has over 40,000 citations on his work, an H-index of 90 (Google Scholar), and has graduated 60 Ph.D. students so far.

Frede Blaabjerg was with ABB-Scandia, Randers, Denmark, from 1987 to 1988. From 1988 to 1992, he got the Ph.D. degree in Electrical Engineering at Aalborg University in 1995. He became an Assistant Professor in 1992, an Associate Professor in 1996, and a Full Professor of power electronics and drives in 1998. From 2017, he became a Villum Investigator. He has published more than 500 journal papers in the fields of power electronics and its applications. He has received 24 IEEE Prize Paper Awards, the IEEE PELS Distinguished Service Award in 2009, the EPE-PEMC Council Award in 2010, the IEEE William E. Newell Power Electronics Award 2014, and the Villum Kann Rasmussen Research Award 2014. He was the Editor-in-Chief of the IEEE TRANSACTIONS ON POWER ELECTRONICS from 2006 to 2012. He is nominated in 2014–2017 by Thomson Reuters to be between the most 250 cited researchers in Engineering in the world. In 2017, he became Honoris Causa at University Politehnica Timisoara (UPT), Romania.

Asad M. Madni served as President, COO & CTO of BEI Technologies Inc. from 1992 until his retirement in 2006. Prior to BEI, he was with Systron Donner Corporation for 18 years in senior technical and executive positions, eventually as Chairman, President, and CEO. He is currently an Independent Consultant, Distinguished Adjunct Professor/Distinguished Scientist at UCLA ECE Department, Faculty Fellow at the UCLA Institute of Transportation Studies, Adjunct Professor at Ryerson University, and Executive Managing Director & CTO of Crocker Capital. He is credited with over 170 publications, 69 issued or pending patents, and is the recipient of numerous national and international honors and awards, including UCSD Gordon Medal for Engineering Leadership, Mahatma Gandhi Pravasi Samman Gold Medal, Ellis Island Medal of Honor, IET J.J.Thomson Medal, IEEE Millennium Medal, TCI College Marconi Medal, UCLA Professional Achievement Medal, UCLA Engineering Alumnus of the Year Award, UCLA Engineering Lifetime Contribution Award, UCLA EE Distinguished Alumni Award, IEEE-HKN Vladimir Karapetoff Award, IEEE AESS Pioneer Award, IEEE HKN Eminent Member Award, IEEE IMS Career Excellence Award, and Tau Beta Pi Distinguished Alumni Award. He is a member of the US National Academy of Engineering, a Fellow of the National Academy of Inventors, a Fellow/Eminent Engineer of 15 of the world's most prestigious professional academies and societies and has been awarded five honorary doctorates and five honorary professorships.

Contributors

Alireza Abbaspour Department of Electrical and Computer Engineering, Florida International University, Miami, FL, USA

Nadia Adnan Department of Management and Humanities, Universiti Teknologi PETRONAS (UTP), Perak, Malaysia

Mohammad Abdullah Al Faruque Department of Electrical and Computer Science, University of California, Irvine, CA, USA

Othman Mohammed Althawadi College of Business and Economics, Qatar University (QU), Doha, Qatar

M. Hadi Amini Carnegie Mellon University, Pittsburgh, PA, USA

Dan Assouline LESO-PB, Ecole Polytechnique Federale de Lausanne, Lausanne, Switzerland

Yvon Besanger G2Elab, Université Grenoble Alpes, CNRS, Grenoble INP, Grenoble, France

Frede Blaabjerg Aalborg University, Aalborg East, Denmark

Kianoosh G. Boroojeni Florida International University, Miami, FL, USA

João P. S. Catalão C-MAST, University of Beira Interior, Covilhã, Portugal; INESC TEC and the Faculty of Engineering of the University of Porto, Porto, Portugal; INESC-ID, Instituto Superior Técnico, University of Lisbon, Lisbon, Portugal

Juan Manuel Corchado BISITE Research Group, University of Salamanca, Salamanca, Spain

Amin Shokri Gazafroudi BISITE Research Group, University of Salamanca, Salamanca, Spain

Pedram Gharani School of Computing and Information, University of Pittsburgh, Pittsburgh, USA

Bakhtyar Hoseinzadeh Department of Energy Technology, Aalborg University, Aalborg, Denmark

S. S. Iyengar Florida International University, Miami, FL, USA

Rupamathi Jaddivada Department of Electrical Engineering and Computer Sciences, Massachusetts Institute of Technology, Cambridge, MA, USA

Saeed Haji Seyed Javadi School of Computing and Information, University of Pittsburgh, Pittsburgh, USA

Shahram Khadivi eBay Inc., Aachen, Germany

Farzad Khorasani Georgia Institute of Technology, Atlanta, GA, USA

Asad M. Madni University of California Los Angeles, Los Angeles, USA

Ricardo D. Medina Instituto de Energía Eléctrica – Universidad Nacional de San Juan, San Juan, Argentina

Sakshi Mishra American Electric Power, Tulsa, OK, USA

Amir Moghadasi Department of Electrical and Computer Engineering, Florida International University (FIU), Miami, FL, USA

Masood Moghaddami Department of Electrical and Computer Engineering, Florida International University (FIU), Miami, FL, USA

Nahid Mohajeri Sustainable Urban Development Programme, Department for Continuing Education, University of Oxford, Oxford, UK

Junita Mohamad-Saleh School of Electrical and Electronic Engineering, Engineering Campus, Universiti Sains Malaysia (USM), Nibong Tebal, Pulau Pinang, Malaysia

Diego X. Morales Smart Grid Research Group, Universidad Católica de Cuenca and Université Grenoble Alpes, CNRS, Grenoble INP, Grenoble, France

Mostafa Rezaei Mozafar Faculty of Engineering, Department of Electrical Engineering, Islamic Azad University, Hamedan, Iran

Soroush Najafi Department of Electrical Engineering, Isfahan University of Technology, Isfahan, Iran

Julie Newman MIT Office of Sustainability, Massachusetts Institute of Technology, Cambridge, MA, USA

Shahrina Md Nordin Department of Management and Humanities, Universiti Teknologi PETRONAS (UTP), Perak, Malaysia

Panos M. Pardalos University of Florida, Gainesville, USA

Imtiaz Parvez Department of Electrical and Computer Engineering, Florida International University (FIU), Miami, FL, USA

Imran Rahman School of Electrical and Electronic Engineering, Engineering Campus, Universiti Sains Malaysia (USM), Nibong Tebal, Pulau Pinang, Malaysia

Amir Safdarian Department of Electrical Engineering, Sharif University of Technology, Tehran, Iran

Anas Salah Eddin Department of Electrical and Computer Engineering, California State Polytechnic University, Pomona, CA, USA

Arman Sargolzaei Department of Electrical and Computer Engineering, Florida Polytechnic University, Lakeland, FL, USA

Arif I. Sarwat Department of Electrical and Computer Engineering, Florida International University (FIU), Miami, FL, USA

Jean-Louis Scartezzini LESO-PB, Ecole Polytechnique Federale de Lausanne, Lausanne, Switzerland

Miadreza Shafie-khah C-MAST, University of Beira Interior, Covilhã, Portugal

Xiazi Sun SYSU-CMU Shunde International Joint Research Institute, Shunde, Guangdong, China

Aditya Sundararajan Department of Electrical and Computer Engineering, Florida International University (FIU), Miami, FL, USA

Saber Talari C-MAST, University of Beira Interior, Covilhã, Portugal

Kang Yen Department of Electrical and Computer Engineering, Florida International University, Miami, FL, USA

Chapter 1
A Panorama of Future Interdependent Networks: From Intelligent Infrastructures to Smart Cities

M. Hadi Amini, Kianoosh G. Boroojeni, S. S. Iyengar,
Frede Blaabjerg, Panos M. Pardalos and Asad M. Madni

1.1 Introduction

There has been an emerging concern regarding the optimal operation of real-world large-scale complex networks [1–4]. In the present volume of this book, as well as the upcoming volumes, we focus on the theory and application of interdependent networks, optimization methods to deal with the computational complexity of such networks, and their role in smart cities. To this end, we consider the influential networks including power and energy networks, transportation networks, wireless sensor networks, communication networks, water networks, and societal networks. Our reliance on these networks as global platforms for sustainable cities and societies has led to the need for developing novel means to deal with arising issues. The considerable scale of such networks brings a large variety of computational complexity and optimization challenges. Although the independent optimization of these networks leads to locally optimum operation points, there is an exigent need

M. H. Amini (✉)
Carnegie Mellon University, Pittsburgh, PA 15213, USA
e-mail: amini@cmu.edu
URL: https://www.HadiAmini.com

K. G. Boroojeni
Florida International University, Miami, USA

S. S. Iyengar
Florida International University, Miami, USA

F. Blaabjerg
Aalborg University, Aalborg East, Denmark

P. M. Pardalos
University of Florida, Gainesville, USA

A. M. Madni
University of California Los Angeles, Los Angeles, USA

© Springer International Publishing AG, part of Springer Nature 2018
M. H. Amini et al. (eds.), *Sustainable Interdependent Networks*,
Studies in Systems, Decision and Control 145,
https://doi.org/10.1007/978-3-319-74412-4_1

to move toward obtaining the globally optimum operation point of such networks while satisfying the constraints of each network properly.

We aim to draw an overview of interdependent networks and their applications in sustainable interdependent networks. Then, we provide a thorough investigation of both classic and modern optimization/control problems in sustainable interdependent networks. We also introduce the physical model and corresponding constraints of these networks elaborately. Further, we aim to study the application of sustainable networks in power microgrids, energy and water networks, and smart cities.

1.2 Smart Cities as Future Interdependent Networks

A smart city is an urban development vision to integrate multiple information and communication technologies in a secure fashion to manage a city's assets including transportation systems, power grids, distributed sensor networks, water supply networks, and other community services [5–8]. Our reliance on these complex networks as global platforms for sustainable cities and societies as well as shortage of global nonrenewable energy sources has raised emerging concerns regarding the optimal and secure operation of these large-scale networks. In a smart city, the data measured by sensor networks in various scales can be utilized to facilitate daily habitants' lives in urban areas, such as using context-aware analysis of sensor data to improve movement of disabled individuals [9], helping law enforcement agents to provide safer environment for all people [10] and measuring people's accessibility to resources and optimize their distributions. The dynamic structure of networks in smart cities, makes us evaluate and study their structure and discover more knowledge about them [11].

1.3 Smart Power Grids and Power Electronic Devices

The electrical power grid is undergoing a transition towards more secure, reliable, sustainable, and efficient system, referred to as smart grid [12–17]. Few large generators are substituted with thousands of smaller distributed power generators (e.g., wind and photovoltaic) which are connected at the distribution line and will in some case provide upstream power to a higher level of voltage. Power electronics by means of power electronic converters is the interconnecting technology to facilitate the power generation to the grid as well as keep the systems synchronized. The technology facilitates also to a large extend controllability in the system like reducing the power production if not needed.

The power electronics interconnection can also be done at much higher voltage in large-scale GW transmission system—either AC or DC and gives the possibility to connect large renewable power production centers placed remotely to concentrations of loads, e.g., large cities [18–21]. The technology is undergoing a steady development with larger and larger power capacity.

In local areas, the multiple generation systems can be organized in micro- or nano-grids—where they are able to operate in islanded mode as well as being connected to an overall grid [22–26]; in such cases, an interlinking power converter is used to facilitate such interconnection and make the transition to the two different modes smooth and efficient. Many efforts are also done to make such grid structures resilient so they make power available for the customer in a sustainable fashion, by deploying efficient computational techniques (e.g., distributed optimization [27, 28], partitioning methods [29], and decentralized techniques [30]), enabling demand side management resources (e.g., residential demand response [31, 32] and flexible electric vehicle charging demand [33–35]). From the computational efficiency perspective, datacenters play a pivotal role in the future complex networks, including smart power girds. They have received significant attention as a cost-effective infrastructure for storing large volumes of data and supporting large-scale services. Dataceners also serve a large number of small and medium sized organizations for their requirements such as financial operations, data analysis, and scientific computations. A key challenge that modern datacenters face is to efficiently support a wide range of tasks (e.g., queries, log analysis, machine learning, graph processing, and stream processing) on their physical platform. The principle bottleneck is often the communication requirement for exchanging the intermediate data pieces among the servers. There has been significant research on flow scheduling algorithms that make better use of network resources. In [36], a low complexity, congestion-aware algorithm is proposed that routes the flows in an online fashion and without splitting.

In [37, 38], dependency among flows is considered as a result of running data parallel applications in data centers, and algorithms with improved performance guarantees are developed.

1.4 Major Aspects of Investigating the Sustainable Interdependent Networks

Here, we provide a general list of different aspects that can be studied to achieve sustainability in future interdependent networks. We aim at covering all these aspects in this book series, including the current volume and the upcoming volumes.

Category A. Classic Optimization and Control Problems consisting of studies from outstanding researchers in the field that discusses and provides an insight into the classical and theoretical problems that exist in optimization and Control.

Category B. Efficient Methods for Optimization Problems in Large-Scale Complex Networks: From Decentralized Methods Toward Fully Distributed

Approaches. A more advanced outlook and possible methodologies that could be incorporated in large networks using various approaches should be highlighted.

Category C. Modeling the Interdependency of Power Systems, Communication Networks, Energy Systems (e.g., Gas, Renewables), Transportation Networks Water Networks, and Societal Networks: The ways in which sustainable interdependence can be achieved across the various networks that span the horizons of smart cities should be explained mathematically in terms of models that could be implemented.

Category D. Application of Power and Communication Networks: Co-simulation Platforms for Microgrids and Communication Networks, and Smart Grid Test Beds: This topic highlights the state-of-the-art technologies that are in place in the area of smart grids.

Category E. Application of Sustainable Interdependent Networks in Future Urban Development: The Tale of Smart Cities: This section constitutes studies that discuss the aspects to be pondered in the future and the direction in which the current school of researchers are heading toward realizing ideal sustained and interdependent networks in the future.

1.5   Overview of Volume I of Sustainable Interdependent Networks

Volume I is divided into four major parts: strategic planning of developing interdependent networks, solutions to performance and security challenges of developing interdependent networks, e-vehicles: a game-changing technology for future of interdependent networks, and promises of power grids for sustainable interdependent networks.

Part 1 starts with Chap. 2 addressing the next-generation sustainability framework at MIT. This promising chapter is authored by Dr. Julie Newman, who joined MIT as the Institute's first Director of Sustainability in the summer of 2013. She has worked in the field of sustainable development and campus sustainability for twenty years. Prior to joining MIT, she was recruited to be the founding Director of the Office of Sustainability for Yale University. Dr. Newman came to Yale from the University of New Hampshire, Office of Sustainability Programs (OSP) where she assisted with the development of the program since its inception in 1997. Prior to her work with the OSP she worked for University Leaders for a Sustainable Future (ULSF). In 2004, she co-founded the Northeast Campus Sustainability Consortium, to advance education and action for sustainable development on university campuses in the northeast and maritime region[1]. This chapter started by calling on a need to "scale and accelerate progress in a manner that leads to organizational

---

[1]Available online, viewed on January 14, 2018. https://dusp.mit.edu/faculty/julie-newman

transformation and measured impact" and recognizes that not every challenge can necessarily be solved the same way at every organization, or in every city.

The next chapter in Part 1 addresses smart cities with special emphasis on energy, communication, data analytics, and transportation. It introduces each of these networks, identifies state-of-the-art in them, and explores open challenges for future research. As its key contribution to the literature, the chapter brings out the interdependencies between these networks through realistic examples and scenarios, identifying the critical need to design, develop, and implement solutions that value such dependencies. One of the introduced interdependencies is the interaction of power and transportation networks. This is an emerging topic which is based on the electrified vehicles and their mutual effect on power systems operation. A comprehensive study on the optimal operation of interdependent power and transportation networks is provided in [39]. Amini et al. [39] proposed a charging station strategy-based routing scheme for electrified vehicles that takes the effect of charging demand on the optimal operating point of power systems into account. There has been an extensive literature on this topic from different perspectives, e.g., optimal allocation of electric vehicle charging station to improve the operation of power systems [40–45], and optimal charge scheduling of electric vehicles considering power system constraints [46–50].

The last chapter of Part 1 is about interdependent interaction of occupational burnout and organizational commitments. It addresses the academic institutions located in Guangxi Province, China, as a case study. This study is based on the review of the related literature and combines the conditions of college in Guangxi. Collecting the data by feasible questionnaire tables among more than 100 college instructors who were selected from about eight colleges of Guangxi, then the data was statistically analyzed by SPSS.

The second part of this volume offers solutions to performance and security challenges of developing interdependent networks in the form of Chaps. 5 and 6. Chapter 5 discussed a few techniques to effectively utilize GPUs in order to process real-world irregular graphs. It first discusses GPU-friendly graph representations such as G-Shards and CW [51] and then introduces warp-efficient methods to dynamically balance the load for GPU's SIMT architecture [52]. Finally, it provides a solution that enables using multiple GPUs to carry out the graph processing task in order to accelerate it more than the previous step [53]. Chapter 6 addresses the security challenges of networked control systems. Chapter 6 summarizes the important security challenges of the networked control systems (NCSs). It covers the related works on the security of NCSs with applications in power, transportation, and unmanned aerial systems. This chapter discusses a new type of attacks to the NCSs called time delay switch attack (TDS), which has been introduced by Sargolzaei for the first time. The current communication protocols and control techniques cannot prevent, detect, and compensate the effects of TDS attacks that could cause much more devastated impacts compared with other types of attacks such as DoS attack due to its smooth nature. A new generalized model is introduced for NCSs under common types of attacks such as false data injection (FDI), time delay switch attack (TDS), and denial of service (DoS) attacks. This chapter will also extend the current techniques to detect FDI and TDS attacks. Then, a resilient communication-control technique is presented and validated to design a load frequency controller for two interconnected power areas under TDS and FDI attacks simultaneously [54, 55]. It is worth noting that the security of complex network is a crucial topic due to the emerging concerns such as cyber-physical attacks [56, 57] and security issues in the smart power grids [58, 59]. Modern and methods are developed to reduce the computation burden of large-scale networks. Among all, oblivious network routing is a promising method that can be deployed to overcome several problems in complex networks [60], such as economic dispatch for power systems [61]. Further, distributed sensor networks have been widely studies in the literature to enhance the observability of several networks, including power systems, transportation networks, and pipelines in gas networks [62].

The third part of this volume examines the role of electric vehicles in the emerging future of sustainable interdependent networks. This part which contains Chaps. 8 and 9 addresses electric vehicles as a game-changing technology. Chapter 8 examines the barriers toward widespread adoption of V2G technology in smart grid environment. Also, Chap. 9 studies the plug-in electric vehicle charging optimization using bio-inspired computational intelligence methods.

The last part of this volume addresses the promises of power grids for sustainable interdependent networks in the form of Chaps. 10–14. Chapter 10 is about coordinated management of residential loads in large-scale systems. Chapter 11 is a methodology review on Estimation of Large-Scale Solar Rooftop PV Potential For Smart Grid Integration. This chapter presents a review on the theory and application of several methods to estimate the solar rooftop PV potential at the national scale. The estimation includes the computation of the following variables: (1) the horizontal solar radiation, (2) the shadowing effects from neighboring buildings and trees over rooftops, (3) the slope, the aspect (direction) distribution, and the shape of rooftops, (4) the solar radiation over the tilted rooftops, and (5) the available rooftop area for PV installation. The reviewed methods include physical and empirical models, geostatistical methods, constant-value methods, sampling methods, geographic information systems (GIS) and light detection and ranging (LiDAR) -based methods, and finally machine learning methods. For more information on this topic please see [63–65]. Chapter 12 studies the optimal SVC Allocation in Power Systems for Loss Minimization and Voltage Deviation Reduction. Finally, Chap. 13 addresses the decentralized control of DR using a multi-agent Method.

Scalable parallel computation platforms play a pivotal role in the computationally-expensive optimization problems of future complex networks. Asghari-Moghaddam et al. investigated the near-data processing (NDP) techniques to provide a scalable parallelism and dramatically reduce the energy consumption. First, it discusses the power consumption in conventional processors and identifies the significant contributors to it. Then, talks about a 3D die-stacking NDP technique to stack an in-order processor on top of each DRAM modules to reduce the data transfer and accelerate the computing while improving the energy efficiency [66]. Finally, a more cost-efficient NDP approach is discussed to use accelerators inside the data buffers of a conventional **LRDIMM** module to reach the same performance of 3D die-stacking technology [67].

In [68], more information on the decentralized control for demand response is provided. Further information about this volume of the book and the upcoming volumes can be found at www.interdependentnetworks.com.

References

1. S. Boccaletti et al., Complex networks: structure and dynamics. Phys. Rep. **424**(4), 175–308 (2006)
2. S.H. Strogatz, Exploring complex networks. Nature **410**(6825), 268–276 (2001)
3. M.E.J. Newman, The structure and function of complex networks. SIAM Rev. **45**(2), 167–256 (2003)
4. J. Duch, A. Arenas, Community detection in complex networks using extremal optimization. Phys. Rev. E **72**(2), 027104 (2005)
5. A. Cocchia, Smart and digital city: a systematic literature review, in *Smart City* (Springer International Publishing, 2014), pp. 13–43
6. R.E. Hall et al., The vision of a smart city. No. BNL–67902; 04042 (Brookhaven National Lab, Upton, NY (US), 2000)
7. P. Lombardi et al., Modelling the smart city performance. Innov. Eur. J. Soc. Sci. Res. **25**(2), 137–149 (2012)
8. H. Arasteh et al., Iot-based smart cities: a survey, in *2016 IEEE 16th International Conference on Environment and Electrical Engineering (EEEIC)* (IEEE, 2016)
9. P. Gharani, H.A. Karimi, Context aware obstacle detection for navigation by visually impaired. Image Vis. Comput. **64** (2017)
10. B. Suffoletto, P. Gharani, T. Chung, H. Karimi, *Using Phone Sensors and an Artificial Neural Network to Detect Gait Changes During Drinking Episodes in the Natural Environment* (Gait Posture, 2017)
11. S.H.S. Javadi, P. Gharani, S. Khadivi, Detecting Community Structure in Dynamic Social Networks Using the Concept of Leadership, arXiv Prepr. arXiv:1711.02053 (2017)
12. H. Farhangi, The path of the smart grid. IEEE Power Energy Mag. **8**(1) (2010)
13. M.H. Amini, B. Nabi, M.-R. Haghifam, Load management using multi-agent systems in smart distribution network, in *Power and Energy Society General Meeting (PES)* (2013)
14. S.M. Amin, B.F. Wollenberg, Toward a smart grid: power delivery for the 21st century. IEEE Power Energy Mag. **3**(5), 34–41 (2005)
15. K.G. Boroojeni, M.H. Amini, S.S. Iyengar, *Smart Grids: Security and Privacy Issues* (Springer International Publishing, 2017)
16. A.I. Sarwat, M. Amini, A. Domijan, A. Damnjanovic, F. Kaleem, Weather-based interruption prediction in the smart grid utilizing chronological data. J. Mod. Power Syst. Clean Energy **4**(2), 308–315 (2016)
17. K.G. Boroojeni, M.H. Amini, S.S. Iyengar, Reliability in smart grids, in *Smart Grids: Security and Privacy Issues* (Springer International Publishing, 2017), pp. 19–29
18. F. Blaabjerg et al., Overview of control and grid synchronization for distributed power generation systems. IEEE Trans. Ind. Electron. **53**(5), 1398–1409 (2006)
19. F. Blaabjerg, Z. Chen, S.B. Kjaer, Power electronics as efficient interface in dispersed power generation systems. IEEE Trans. Power Electron. **19**(5), 1184–1194 (2004)
20. F. Blaabjerg, M. Liserre, K. Ma, Power electronics converters for wind turbine systems. IEEE Trans. Ind. Appl. **48**(2), 708–719 (2012)
21. S.B. Kjaer, J.K. Pedersen, F. Blaabjerg, A review of single-phase grid-connected inverters for photovoltaic modules. IEEE Trans. Ind. Appl. **41**(5), 1292–1306 (2005)

22. R.H. Lasseter, P. Paigi, Microgrid: a conceptual solution, in *2004 IEEE 35th Annual Power Electronics Specialists Conference, 2004. PESC 04*, vol. 6 (IEEE, 2004)
23. S. Parhizi et al., State of the art in research on microgrids: a review. IEEE Access **3**, 890–925 (2015)
24. M.H. Amini, K.G. Boroojeni, T. Dragičević, A. Nejadpak, S.S. Iyengar, F. Blaabjerg, A comprehensive cloud-based real-time simulation framework for oblivious power routing in clusters of DC microgrids, in *2017 IEEE Second International Conference on DC Microgrids (ICDCM)* (IEEE, 2017)
25. A.L. Dimeas, N.D. Hatziargyriou, Operation of a multiagent system for microgrid control. IEEE Trans. Power Syst. **20**(3), 1447–1455 (2005)
26. M.H. Amini, K.G. Broojeni, T. Dragičević, A. Nejadpak, S.S. Iyengar, F. Blaabjerg, Application of cloud computing in power routing for clusters of microgrids using oblivious network routing algorithm, in *19th European Conference on Power Electronics and Applications (EPE'17 ECCE Europe)* (2017)
27. J. Mohammadi, G. Hug, S. Kar, Agent-based distributed security constrained optimal power flow. IEEE Trans. Smart Grid (2016)
28. M.H. Amini, R. Jaddivada, S. Mishra, O. Karabasoglu, Distributed security constrained economic dispatch, in *IEEE Innovative Smart Grid Technologies-Asia (ISGT ASIA)* (IEEE, 2015), pp. 1–6
29. M. Mehrtash, A. Kargarian. A. Mohammadi, A partitioning-based bus renumbering effect on interior point-based OPF solution, in *Power and Energy Conference (TPEC)* (IEEE Texas, IEEE, 2018)
30. A. Mohammadi, M. Mehrtash, A. Kargarian, Diagonal quadratic approximation for decentralized collaborative TSO+DSO optimal power flow. IEEE Trans. Smart Grid (2018). https://doi.org/10.1109/TSG.2018.2796034
31. Z. Chen, L. Wu, Y. Fu, Real-time price-based demand response management for residential appliances via stochastic optimization and robust optimization. IEEE Trans. Smart Grid **3**(4), 1822–1831 (2012)
32. M.H. Amini, J. Frye, M.D. Ilic, O. Karabasoglu, Smart residential energy scheduling utilizing two stage mixed integer linear programming, in *North American Power Symposium (NAPS), 2015* (IEEE, 2015), pp. 1–6
33. S. Shao, M. Pipattanasomporn, S. Rahman, Grid integration of electric vehicles and demand response with customer choice. IEEE Trans. Smart Grid **3**(1), 543–550 (2012)
34. S. Shao, M. Pipattanasomporn, S. Rahman, Demand response as a load shaping tool in an intelligent grid with electric vehicles. IEEE Trans. Smart Grid **2**(4), 624–631 (2011)
35. M.H. Amini, M.P. Moghaddam, Probabilistic modelling of electric vehicles' parking lots charging demand, in *2013 21st Iranian Conference on Electrical Engineering (ICEE)* (IEEE, 2013), pp. 1–4
36. M. Shafiee, J. Ghaderi, A simple congestion-aware algorithm for load balancing in datacenter networks. IEEE/ACM Trans. Netw. **25**(6), 3670–3682 (2017)
37. M. Shafiee, J. Ghaderi, Scheduling coflows in datacenter networks: improved bound for total weighted completion time, in *Proceedings of the 2017 ACM SIGMETRICS/International Conference on Measurement and Modeling of Computer Systems* (ACM, 2017)
38. M. Shafiee, J. Ghaderi, Brief announcement: a new improved bound for coflow scheduling, in *Proceedings of the 29th ACM Symposium on Parallelism in Algorithms and Architectures* (ACM, 2017)
39. M.H. Amini, O. Karabasoglu, Optimal operation of interdependent power systems and electrified transportation networks. Energies **11**, 196 (2018)
40. M.H. Amini, M.P. Moghaddam, O. Karabasoglu, Simultaneous allocation of electric vehicles' parking lots and distributed renewable resources in smart power distribution networks. Sustain. Cities Soc. **28**, 332–342 (2017)
41. M.H. Amini, A. Islam, Allocation of electric vehicles' parking lots in distribution network, in *Innovative Smart Grid Technologies Conference (ISGT), 2014 IEEE PES* (IEEE, 2014), pp. 1–5

42. A.S.B. Humayd, K. Bhattacharya, Distribution system planning to accommodate distributed energy resources and PEVs. Electr. Power Syst. Res. **145**, 1–11 (2017)
43. Y. Li, P. Zhang, Y. Wu, Public recharging infrastructure location strategy for promoting electric vehicles: a bi-level programming approach. J. Clean. Prod. **172**, 2720–2734 (2018)
44. M.R. Mozafar, M.H. Moradi, M.H. Amini, A simultaneous approach for optimal allocation of renewable energy sources and electric vehicle charging stations in smart grids based on improved GA-PSO algorithm. Sustain. Cities Soc. **32**, 627–637 (2017)
45. M.H. Amini, P. McNamara, P. Weng, O. Karabasoglu, Y. Xu, Hierarchical electric vehicle charging aggregator strategy using Dantzig-Wolfe decomposition. IEEE Design Test Mag. (2017)
46. H. Zhang et al., Joint Planning of PEV Fast-Charging Network and Distributed PV Generation. arXiv preprint arXiv:1702.07120 (2017)
47. M.R. Mozafar, M.H. Amini, M.H. Moradi, Innovative appraisement of smart grid operation considering large-scale integration of electric vehicles enabling V2G and G2V systems. Electr. Power Syst. Res. **154**, 245–256 (2018)
48. M.H. Amini, A. Kargarian, O. Karabasoglu, ARIMA-based decoupled time series forecasting of electric vehicle charging demand for stochastic power system operation. Electr. Power Syst. Res. **140**, 378–390 (2016)
49. M.H. Amini, O. Karabasoglu, M.D. Ilić, K.G. Boroojeni, S.S. Iyengar, Arima-based demand forecasting method considering probabilistic model of electric vehicles' parking lots, in *Power & Energy Society General Meeting, 2015 IEEE* (IEEE, 2015), pp. 1–5
50. N. Adnan et al., Adoption of plug-in hybrid electric vehicle among Malaysian consumers. Ind. Eng. Manage **5**(185), 0316–2169 (2016)
51. F. Khorasani, K. Vora, R. Gupta, L.N. Bhuyan, CuSha: vertex-centric graph processing on GPUs, in *Proceedings of the 23rd International Symposium on High-Performance Parallel and Distributed Computing* (ACM, June 2014), pp. 239–252
52. F. Khorasani, K. Vora, R. Gupta, L.N. Bhuyan, Enabling work-efficiency for high performance vertex-centric graph analytics on GPUs, in *Proceedings of the Seventh Workshop on Irregular Applications: Architectures and Algorithms (IA3'17)* (ACM, New York, 2017). https://doi.org/10.1145/3149704.3149762
53. F. Khorasani, High performance vertex-centric graph analytics on GPUs. PhD diss., University of California, Riverside (2016)
54. A. Sargolzaei et al., Resilient design of networked control systems under time delay switch attacks, application in smart grid. IEEE Access **5**, 15901–15912 (2017)
55. A. Sargolzaei, K.K. Yen, M.N. Abdelghani, Preventing time-delay switch attack on load frequency control in distributed power systems. IEEE Trans. Smart Grid **7**(2), 1176–1185 (2016)
56. F. Pasqualetti, F. Dörfler, F. Bullo, Attack detection and identification in cyber-physical systems. IEEE Trans. Autom. Control **58**(11), 2715–2729 (2013)
57. F. Pasqualetti, F. Dörfler, F. Bullo, Cyber-physical attacks in power networks: models, fundamental limitations and monitor design, in *2011 50th IEEE Conference on Decision and Control and European Control Conference (CDC-ECC)* (IEEE, 2011)
58. P. McDaniel, M. Stephen, Security and privacy challenges in the smart grid. *IEEE Security & Privacy* 7.3 (2009)
59. K.G. Boroojeni, M. Hadi Amini, S.S. Iyengar, Overview of the security and privacy issues in smart grids, in *Smart Grids: Security and Privacy Issues* (Springer International Publishing, 2017), pp. 1–16
60. S.S. Iyengar, K.G. Boroojeni, *Oblivious Network Routing: Algorithms and Applications* (MIT Press, 2015)
61. K.G. Boroojeni et al., An economic dispatch algorithm for congestion management of smart power networks. Energy Syst. **8**(3), 643–667 (2017)
62. S.S. Iyengar, K.G. Boroojeni, N. Balakrishnan, *Mathematical Theories of Distributed Sensor Networks* (Springer, 2014)

63. N. Mohajeri, D. Assouline, B. Guiboud, A. Bill, A. Gudmundsson, J.L. Scartezzini, A city-scale roof shape classification using machine learning for solar energy applications. Renew. Energy **121**, 81–93 (2018)
64. D. Assouline, N. Mohajeri, J.L. Scartezzini, Building rooftop classification using random forests for large-scale PV deployment, in *Earth Resources and Environmental Remote Sensing/GIS Applications VIII* (International Society for Optics and Photonics, 2017), vol. 10428, p. 1042806
65. D. Assouline, N. Mohajeri, J.L. Scartezzini, Quantifying rooftop photovoltaic solar energy potential: a machine learning approach. Solar Energy **141**, 278–296 (2017)
66. H. Asghari-Moghaddam, A. Farmahini-Farahani, K. Morrow, J.H. Ahn, N.S. Kim, Near-DRAM acceleration with single-ISA heterogeneous processing in standard memory modules. IEEE Micro **36**(1), 24–34 (2016)
67. H. Asghari-Moghaddam, Y.H. Son, J.H. Ahn, N.S. Kim, Chameleon: versatile and practical near-DRAM acceleration architecture for large memory systems, in *2016 49th Annual IEEE/ ACM International Symposium on Microarchitecture (MICRO)* (IEEE, 2016), pp. 1–13
68. S. Lu et al., Centralized and decentralized control for demand response. *Innovative Smart Grid Technologies (ISGT), 2011 IEEE PES* (IEEE, 2011)

Part I
Strategic Planning of Developing Sustainable Interdependent Networks

Chapter 2
Calling for a Next-Generation Sustainability Framework at MIT

Julie Newman

2.1 Why Higher Education

Higher education has often been a catalyst in driving both national and global innovation and sweeping shifts in cultural norms. Many college and university campuses have committed to and have demonstrated leadership in the global sustainability movement over the past two decades by establishing sustainability offices and officer positions, setting predominantly achievable operational and academic targets and tracking various metrics to demonstrate progress towards a "state of sustainability", over time. In some instances, aspirational goals such as carbon neutrality or zero waste have been established as well, which signal long-term direction setting and a willingness to grapple with complexity. These are all points on the trajectory of progress in a two-decade time horizon; however, the time has come to explore how to scale and accelerate progress in a manner that leads to organizational transformation and measured impact—locally to globally. This is not a challenge that can be automated, mechanized or even imposed. Nor is it a challenge that can necessarily be solved the same way at every organization, or in every city. Our ability to measure the collective impact of higher education in advancing sustainability globally has yet to be readily quantified. That said, I argue that there is a knowledge base, combined with organizational change management strategies, processes and aspirations to drive a commitment to sustainability that can be shared across organizations grounded in interdependent systems regardless if the implementation plan is different from one organization to another.

Organizations, such as higher education, understand that one organization alone will not change the trajectory on climate change, improve air quality, reduce single

J. Newman (✉)
MIT Office of Sustainability, Massachusetts Institute
of Technology, Cambridge, MA, USA
e-mail: j_newman@mit.edu

© Springer International Publishing AG, part of Springer Nature 2018
M. H. Amini et al. (eds.), *Sustainable Interdependent Networks*,
Studies in Systems, Decision and Control 145,
https://doi.org/10.1007/978-3-319-74412-4_2

occupancy vehicle dependency or shrink the waste stream. However, the sum of the parts is the aggregate of the research emerging from our laboratories, the educational outcomes of our graduates and the design and operation of our campuses. Twenty years into the emergence of the role of higher education advancing and solving for sustainable development strategies, processes and solutions, we have a strong understanding as to what can be foundationally implemented across our campus systems and shared with other organizations. Yet, we seem to be challenged by sharing and readily adapting best practices. Moreover, cities, states and provinces are made of organizations all reliant upon shared systems and services that provide or enable access to energy, water, mobility, infrastructure, food, goods, delivery systems, and waste removal. Each of these same organizations is vulnerable should one of these systems fail, underperform, incur deferred maintenance. Equally each of these organizations and their users benefits when these shared systems are thriving and optimized. The interactions between human health, environmental quality and fiscal responsibility are increasingly complex. Even routine decisions become complicated when we take into account the interconnectedness of the systems that sustain us.

This chapter will explore and provide insight into how to establish an Office of Sustainability from inception to implementation via a case example from the Massachusetts Institute of Technology, Office of Sustainability.

2.1.1 Launching an Office of Sustainability

The success of a Sustainability Officer and the office they manage is reliant upon a comprehensive understanding of the organization [including the priorities, financial structures, hierarchy, value system] and the systems [political, municipal, infrastructure] in which the organization is embedded. If one does not have a comprehensive understanding of both contexts, then the resulting approaches will be incremental at best, marginal in nature and not lead to the foundation for organizational transformation needed to become sustainable. Secondly, a Sustainability Officer relies upon the ability and willingness to build, facilitate, and manage multilateral relationships and processes and embody an agile approach as budgets fluctuate and leadership priorities evolve. One of the most effective approaches to building an office be it within a university or company is to ensure that the office structure, value system and approach is "of the place" and not "attached to". Long-term change management for sustainability must take place from within.

2.2 Setting the Stage at MIT

The Massachusetts Institute of Technology [MIT] is one of 58 colleges and universities within the Metropolitan Boston area which collectively enrol >250,000 students annually. In 2013, the late Mayor Menino launched an initiative, now supported by the current Mayor, Marty Walsh, entitled the Boston Green Ribbon Commission, an advisory committee with leadership representation from Harvard, MIT, North-eastern, UMass Boston and Boston University alongside the CEO's from Banks, NGOs such as CERES and the local utility companies. All at the table due to a common concern related to the impacts of climate change on our city, our region, our economy and our population.

Cambridge, where MIT is located, boasts a community fabric that includes an innovation district, within a city that demonstrates robust civic engagement and municipal leadership sustainability. In 2013, MIT co-signed the Cambridge Community Compact for a Sustainable Future with the leadership of Harvard and the Mayor at that time. Today there is a commitment to seek Net Zero emission across the building sector by 2035 and several other initiatives, ranging from an exploration of low carbon energy sourcing to the development of a community carbon fund that provides a robust municipal context for sustainability.

Recent climate vulnerability analysis lead, independently, by the cities of Boston and Cambridge denotes the challenges we share and thus the need to seek joint solutions over time.

In addition to optimizing interdependent infrastructure systems that enable resiliency, our partnerships with the Cities of Cambridge and metro Boston, and regional and global sustainability networks provide further opportunity to advance collective intelligence around shared problems that affect our common resources and mutual desire for a more sustainable future. Each critical decision we make on campus is an opportunity to think deeply about how to ensure the growth of human potential while maintaining the integrity of the natural systems that support life on the planet. A commitment to sustainability enables a process that calls upon us to grapple with how best to weigh fiscal, environmental and human health impacts and at varying timescales. Within MIT, we are now building the capacity to create and sustain this process over time, and in partnership with Cambridge and Boston where feasible and essential.

2.3   Changing the Game for Campus Sustainability

MIT is positioned to excel in a next-generation approach to campus sustainability. With an active administration, emphasis on sustainable guidelines in planning, campus construction, energy performance, an incentivized alternative mobility system, and a collaborative fabric of researchers, entrepreneurs, inventors, and community partners, we are now poised to both accelerate our successful performance and provide leadership on the meaning of creating a "sustainable campus". At MIT, we have a particularly unique opportunity to transform the campus into a test bed for change, frequently referred to as a living laboratory and one, which matches cutting edge sustainable urban strategies emerging from our classrooms and laboratories.

At MIT, we are developing a model of campus sustainability that advances the educational mission of the institute. If we are teaching through demonstration that time and budget [in their current construct] keeps us from solving complex and harmful challenges, then we are not standing up to our mission. Herein lies a challenge of our time. We believe that MIT is poised to stand up to the enormity of this task.

At MIT, our goal is to advance a next-generation campus sustainability model that both advances the mission of academic excellence at the institute and transforms our campus in yet another demonstration of MIT's ability to marshal knowledge in addressing the world's great challenges. In this context, the mission of the office is to transform MIT into a powerful model that generates new and proven ways of responding to the unprecedented challenges of a changing planet via operational excellence, education, research and innovation on our campus.

This now leads me to a discussion of our model. We aim to transform MIT into a powerful model that generates new and proven ways of responding to the unprecedented challenges of a changing planet via operational excellence, education, research and innovation on our campus. A set of values has been articulated and embraced that combines an understanding of the culture of MIT and what is needed to advance sustainability on and off campus: applied innovation, collective intelligence, civic responsibility and systems thinking.

We seek a model that considers scales of impact and define this as stated here: to inspire and enable the continuous generation of breakthrough sustainability solutions to transform our campus, city and the globe. We want people to find their place in this framework and recognize the deep interconnectivity—and need for working across systems and scales.

The areas of responsibility for the office are organized as follows: sustainable campus systems, urban living and learning laboratory, leadership and capacity building and collaborative partnerships. In the start-up phase of the office, the foundation is built on developing an understanding of strategy for building sustainable campus systems and forging strategic collaborative partnerships within and external to the institute. In the end, our success grounded in the framework outlined here relies upon how we work. An Office of Sustainability continuously connects people, ideas, data and systems in ways that inspire transformative and lasting change. This can take time (Fig. 2.1).

In the end, our goal is to inspire and enable the continuous generation of breakthrough sustainability solutions and systems to transform our *campus, city and globe*. By framing the commitment to sustainability in this context, we provide a filter by which all assessments, solutions and even strategic partnership must be vetted. When applied successfully, a university and the solutions sought therein are interconnected via a sustainable **interdependent network of systems, services, processes and political structures advancing common goals and objectives**. I propose that this approach is the foundation needed to launch a next-generation framework for sustainability. A core component of which is the creation of a decision-making platform for the campus that is founded on systems thinking and powered by responsive, robust data.

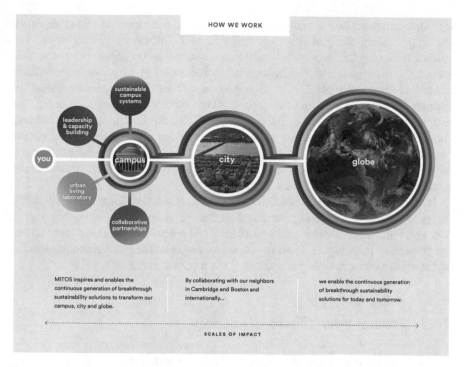

Fig. 2.1 MIT Office of Sustainability framework

2.4 Game-Changing Strategies with Measurable Results

In an effort to highlight the application of this framework, I will focus on three unique projects that are now being implemented. The projects span the process of adapting a global climate vulnerability model to the campus scale to a partnership-based power purchase agreement for large-scale solar to provide free access to public in an effort to reduce single occupancy vehicles to campus by 10% in two years. The common thread is that each of these is system-based and measurable.

2.5 Climate Leadership

MIT's commitment to researching, mitigating and educating about climate change is robust. In the fall of 2014, MIT held a community-wide conversation on climate called for by the President of the Institute. This catalyzed a ground-up dialogue across all schools and departments, including faculty, staff and students as to how MIT should shape its commitment to leadership on climate change. A year later in the fall of 2015, the *MIT Plan for Action on Climate Change* was published and articulated an unequivocal position that climate change is one of the most important challenges facing our global society and that MIT will take on many of the scientific, technological and social challenges needed to contribute to the understanding and mitigation of the impacts of global warming. Organized around five pillars[1] of focus, the plan seeks to align the full array of expertise within the Institute to advance solutions not only through academic research, teaching and learning, but also in practice. MIT recognizes that finding solutions to climate change requires interdisciplinary approaches across the Institute that mobilizes not only MIT's research and educational resources, but also operational practices to have maximum impact. This in and of itself was game changing for the institute and leveraged MIT as a system not a unit.

In the light of this commitment, the MIT Office of Sustainability was responsible for simultaneously managing the development of a mitigation approach with a focus on reducing emissions and a resiliency and adaptation planning process. For the purpose of this chapter, I will focus on two projects: climate resiliency planning and mitigating greenhouse gas emissions via a power purchase agreement.

2.5.1 Planning for Climate Resiliency

The US National Climate Assessment (NCA) report, *Global Climate Change Impacts in the USA*, reports that communities in the USA are experiencing impacts from climatic changes, including more frequent and extreme precipitation and heat events, rising sea levels and insecurities to regional supply systems. These impacts have the potential to interrupt the operations and mission of MIT through the disruption of daily activities, energy supply to campus, and rising costs as a result of changing "norms" for building operation and land and water management. For MIT to create a resiliency and adaptation plan, the campus first needed to complete a *vulnerability assessment* of the potential climate impacts and their corresponding severity.

It is these strategies which distinguish the sustainability leadership model for higher education from the private sector: our researchers and laboratories. The Office of Sustainability established a partnership with researchers from the Joint Program on Science and Policy at MIT and brought them together with members of the Departments of Facilities, Planning, Utilities, Insurance, and Emergency Management to develop a problem definition and methodology. From this initial

---

[1]
- Pillar A: Improve MIT's understanding of climate change and advance novel, targeted mitigation and adaptation solutions
- Pillar B: Accelerate progress towards low- and zero-carbon energy technologies
- Pillar C: Educate a new generation of climate, energy and environmental innovators
- Pillar D: Share what MIT knows and learn from others around the world
- Pillar E: Use MIT's community as a test bed for change.

Fig. 2.2 Summit farms 60 MW solar array, North Carolina

process, a multi-stakeholder, multi-phase analysis was developed that has lead to the completion of a phase 1 flood vulnerability study at the time of this writing. The role of the Office of Sustainability is to facilitate this interdependent network of players and data in phase 1 and the integration of these findings with city vulnerability analysis.

Climate vulnerability cannot be assessed merely within the bounds of a campus but rather has to be integrated into municipal context in order to truly understand and plan for the future campus (Fig. 2.2).

2.6 Power Purchase Agreement

MIT has embraced the challenge outlined in the *Plan for Action on Climate Change* to use the campus as test bed for experimentation and innovation in reducing its GHG emissions by a minimum of 32% by 2030 from a 2014 baseline and aspire to being carbon neutral as soon as possible. In addition to the process of reducing on-site emissions, MIT has set out to explore how to reduce emissions via national opportunities and within a different, potentially dirtier, grid. In the light of this, MIT advanced our commitment to climate mitigation by joining with two local partners [Boston Medical Center and Post Office Square Redevelopment Corporation] in a twenty-five-year power purchase agreement with Dominion Resources. This agreement not only demonstrates the value of aggregation but will provide MIT researchers with first-hand access to data from the 60 MW solar farm that has 255,000 solar panels on approximately 650 acres of land. However, there is not a clear path for large, energy-intensive research institutions like MIT to meet the necessary GHG reduction levels nor sufficient understanding of the optimal balance of new technology deployment, energy efficiency investments, and on-site and off-site renewable energy procurement.

The game-changing attributes in this case are twofold: the formal tie between the solar farm, MIT researchers [including faculty and students], Dominion Energy and the opportunity to leverage the campus as a test bed for solar. In summary, we recognize that there is an unprecedented opportunity to create a solar test bed to de-risk emerging solar technologies, proving their viability to financiers and developers that could have an impact nationally and globally. MIT has seized the opportunity to demonstrate leadership in campus greenhouse gas reduction strategies and seeks to amplify these impacts in time.

2.7 Transforming Systems

This chapter started by calling on a need to "scale and accelerate progress in a manner that leads to organizational transformation and measured impact" and recognizes that not every challenge can necessarily be solved the same way at every organization, or in every city. That said, I call upon us to make our knowledge base accessible. The underlying characteristics of a process that calls upon a multi-stakeholder approach, informed by robust data and facilitated by an office/person able to drive a systems thinking solutions framework is at the heart of the success of MIT's accelerated integration and implementation of sustainability principles across the institute.

Chapter 3
Toward a Smart City of Interdependent Critical Infrastructure Networks

Arif I. Sarwat, Aditya Sundararajan, Imtiaz Parvez, Masood Moghaddami and Amir Moghadasi

3.1 Introduction

With an increasing deployment of distributed intelligent devices that engage in ubiquitous data acquisition and pervasive computing, our way of living is definitely bound to undergo major paradigm shifts. One of the key ideas that promote sustainable future with increased quality of living is a smart city.

3.1.1 Existing Literature and Motivation

Every city has different infrastructure that the consumers expect to be available on demand, reliable to operate, resilient to failures, robust to attacks, and scalable to support changing needs and population demographics. These infrastructures, referred to interchangeably as *networks* in this chapter, are the power grid, transportation, waterway, communication and analytics, oil and gas, consumer services, emergency

A. I. Sarwat (✉) · A. Sundararajan · I. Parvez
M. Moghaddami · A. Moghadasi
Department of Electrical and Computer Engineering,
Florida International University (FIU), Miami, FL, USA
e-mail: asarwat@fiu.edu

A. Sundararajan
e-mail: asund005@fiu.edu

I. Parvez
e-mail: iparv001@fiu.edu

M. Moghaddami
e-mail: mmogh003@fiu.edu

A. Moghadasi
e-mail: amogh004@fiu.edu

© Springer International Publishing AG, part of Springer Nature 2018
M. H. Amini et al. (eds.), *Sustainable Interdependent Networks*,
Studies in Systems, Decision and Control 145,
https://doi.org/10.1007/978-3-319-74412-4_3

services like health care and fire, and societal management services to improve quality of life and well-being among citizens. In today's city, these networks are treated independent of one another, with little to no interaction between their constituent elements. However, it is bound to change in the model offered by a city that is heavily interconnected and *smart*.

In a smart city, these networks engage in complex interactions and information exchange via a flexible communication infrastructure, augmented by high-end data processing, optimization, control and visualization models. The intelligence of these networks is more decentralized and people-inclusive. The Internet of Things (IoT) becomes part of a broader Internet of Everything (IoE) paradigm which also includes policymaking, human behavior, decision making, and social analytics.

Consider a scenario to illustrate how smart city is an interdependent network of things and people. Each household uses Electric Vehicles (EVs) for commutation. Being smart loads that are spontaneous and indispensable, required energy must be available either in reserve (through energy storage technologies) or on demand. These EVs further have the capability to trade energy via grid-to-vehicle (G2V), vehicle-to-grid (V2G) or vehicle-to-vehicle (V2V) technologies. Each consumer has access to an application that constantly collects data from their EVs and reports information about performance, vehicular quality, state of charge (SoC), and a map of nearest charging stations with waiting times. This application requires a resilient and pervasive communication infrastructure coupled with high-speed distributed data analytics. Occasionally, the application avails cloud services. Further, it can be personalized by each consumer, reflecting their requirements on maintenance and operation plans compliant with jurisdictional policies. Alongside, a command and control center conducts traffic optimization and congestion control and monitors the state of roadways for real-time situation awareness using powerful visualization and cloud analytics.

Another scenario that highlights the need for sustainability of smart city networks involves emergency services like health care and how they are interlinked with other networks. With the medical industry increasingly using advanced data processing tools to manage patient information and provide medical services, its reliance on analytics and communication is only expected to increase in a smart city environment. Autonomous ambulances could exist which must navigate along the shortest routes to reach patients in emergencies. Determining such routes is contingent on various constraints, for which the use of optimization is imperative. Electrified public shuttles could shuttle between communities and nearest hospitals for senior citizens to go on their regular visits. An interruption in medical services due to power outages can be expensive. Thus, this calls for frequent interactions with the energy infrastructure as well. It can be seen that under such comprehensive scenarios, a smart city is no longer a mere coexistence of networks, but also one that leverages their interdependency through constant interactions between them [1].

To present the concepts clearly, this chapter lays special emphasis on four networks that are fundamental and indispensable to a smart city environment: communication, data analytics, energy, and transportation. Some or most of the challenges specified for these networks could also be applicable to other networks, but further

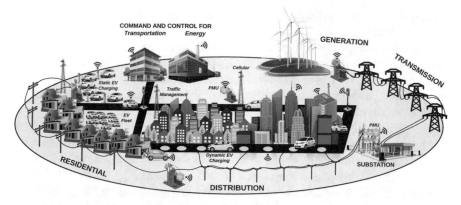

Fig. 3.1 A conceptual smart city with energy, transportation, communication and analytics

discussion of them is beyond the scope of this chapter. A conceptual layout of future smart city with these four networks is illustrated by Fig. 3.1.

The interconnectivity between smart city networks is not a distant vision but an imminent reality, with significant signs already evident [2–7]. To ensure they are also sustainable, the networks must be constantly optimized, secured, made reliable and flexible to the changing needs and demographics of prosumers (consumers who also have the capability to produce services). There is a rampant need for such models in each of the constituent city networks. Some examples where optimization, control and visualization play crucial roles are in determining the power grid states for most reliable operation, cheapest ways to power homes and offices, most efficient way to harness renewable energy sources (RESs) through distributed control, dispatch and commit the generation units that yield lowest costs, manage and route traffic in roadways that offers the least commutation time between two frequented locations during peak hours, schedule and manage processing tasks at a data center to ensure lowest response time, minimized processor power consumption or minimized computational complexity, and route large volumes of data packets over a potentially congested network to minimize the cost of transmission, and reduce the likelihoods of propagation delay or packet loss. To establish and sustain smart cities by the year 2030, a holistic vision to effectively integrate, operate, and manage these networks is required.

A few recent works in the literature consider the interdependency between some of the smart city networks. An oblivious network routing design powered by a Platform as a Service (PaaS) cloud computing platform was proposed for power routing between clusters of microgrids [8]. This work takes a significant step by seamlessly integrating the three smart city elements: energy, communication, and computation. It handles the challenge of scalability by emulating the system over a real-time digital simulator (RTDS) and conducts an evaluation of the bidirectional communication between the cloud and the individual microgrids using OMNeT++. Challenges in interoperability while integrating with legacy systems of the power grid and associated cyber-physical security concerns, however, were not emphasized. Keeping the reliability and stability of smart grid at the center, security and privacy concerns were elaborated and mitigation solutions were recommended [9]. This work accounted for the interdependency between power system, communication, and end-users, but laid much of its emphasis on the cloud networks. With the network intelligence getting decentralized, the emergence of fog computing is expected to reduce the strain on the cloud but widen the attack surface. It also mandates the grid's edge devices to be equipped with a certain level of attack-preparedness they now lack. Another work tackles a specific issue within the power system networks: interruptions in reliable power delivery due to inclement weather conditions [10]. It is well known that reliability of power system is directly influenced by weather. This work attempts to predict weather-induced interruptions using time-series historical data to train an artificial neural network (ANN) model and validates its predictions against the actual interruption data collected from the same localities. This study indirectly attempts to improve the quality of life for consumers in a given locality. A similar study assessed the smart grid reliability due to changes in weather conditions using a Boolean-Driven Markov Process, but considers it specifically under grid-tied Distribution Generation (DG) scenarios [11]. Significantly, the paper arms the energy network of future smart city with an effective way to conduct reliability assessment under RES-integration on the demand side, since conventional indices to assess system reliability would no longer be applicable. In the context of interdependent networks, there is a correlation between reliability and quality of life for consumers and system stability. Hence, a more effective way to assess reliability implies better availability of the power supplied, which is a foundational aspect of many other city networks of the future like communication, analytics, and transportation. With a brief discussion of the chapter's motivation and related work in the literature, it is important to provide the key contributions.

3.1.2 Contribution of the Chapter

The binding quality of smart city networks is that they are logically independent but functionally interdependent on each other. Although it is widely recognized that smart city comprises different interdependent networks, the solutions developed to address the key challenges elicited earlier still exclusively consider the specific network they are designed for. This chapter aims to bridge this gap in research by providing a roadmap to the nature of solutions required for managing and optimizing these networks.

Besides identifying the state of the art in each of the four networks and briefly summarizing the key challenges to future research in them, the chapter also illumines the underlying interdependency between them through realistic scenarios, thereby justifying why the sustainability of smart city networks is as challenging as it is important for the seamless operation and management of future smart cities. With this precursory knowledge, future researchers in this cross-functional domain will be empowered to develop solutions that are not only intelligent, but also inclusive of societal aspects, interoperability, scalability, and associated dependencies. Having described the chapter's key contributions, its organization is summarized below.

3.1.3   Organization of the Chapter

This chapter aims to address the challenge of interdependent smart city networks by introducing and describing intelligent communication, decentralized data analytics, smart energy design, and smart transportation for future city scenarios, including state of the art and key open research challenges. Further, it elaborates on some of these challenges through realistic examples and scenarios to signify and emphasize their importance in the context of smart cities. The remainder of this chapter is organized as follows. Section 3.2 presents the architecture and functionalities of the backbone communication while Sect. 3.3 highlights the paradigm shift from centralized cloud-based analytics to decentralized fog-oriented approach where the future of big data processing lies. Section 3.4 delves into the integration of RESs, existing research and methodologies. Section 3.5 introduces the key concepts and state-of-the-art elements of a smart transportation infrastructure. Each section ends with a brief insight into important challenges and future research directions to provide a guiding start for the readers to conduct further studies. Section 3.6 summarizes the chapter and issues concluding remarks from the authors.

3.2   Intelligent Communication

Communication network has the same level of significance in a smart city that the cardiovascular system has in a human body. The principles of IoT are neatly encapsulated by a communication infrastructure, with diverse devices spread across the vast geographical landscape of the cities, constantly sensing, reporting, and acting upon data. They typically employ wireless or wired communication media to transmit and receive data. Advanced metering infrastructure (AMI) smart meters for electricity and water, wide area monitoring, protection, automation and control (WAMPAC) devices like synchrophasors, vehicle-to-vehicle (V2V), grid-to-vehicle (G2V), and vehicle-to-grid (V2G) applications, traffic control and management systems, smart inverters and weather stations, smart buildings, and other intelligent electronic devices (IEDs) for protection and control are only some of the many devices that require ubiquitous and pervasive communication support to perform their intended functions effectively [12]. These devices engage in different forms of communication, ranging from machine-to-machine (M2M), machine-to-human (M2H), and machine-to-software (M2S). The future smart city is envisaged to support peer-to-peer bidirectional information exchange where the factors of trust and mutual authentication will gain prominence. To achieve this, there exist different protocols such as LTE-U, WiFi and ZigBee for wireless communication, and power line communication (PLC) for wired.

Consider an example of V2V applications, where Electric Vehicles (EVs) are required to engage in direct peer-to-peer communication to make important decisions about traffic congestion, collision avoidance, better platooning, and autonomous driving. In this case, dedicated short-range communication (DSRC) and WiFi are envisioned as the best candidate solutions. For the scenario of a smart home building where different appliances like smart TV, dishwasher, laundry systems, and heating, ventilation, and air conditioning (HVAC) systems constantly interact with the building's designated AMI smart meter, in which WiFi, Bluetooth, and ZigBee turn out to be the accepted communication protocols. For realizing meter-to-meter or meter-to-access point (AP) communications, long-range protocols like LTE/LTE-U may be exploited. In systems where a large topography of IoT sensors are interconnected, the use of Narrowband Internet of Things (NBIoT) might be considered feasible for monitoring and data collection.

Therefore,the communication requirement of smart cities can be long, medium and short ranged [13–15]. For home appliances, smart devices, local sensors, V2V communication, metering infrastructure, etc., short- and medium-range communication is required. On the other hand, personal mobile communication, Internet/IoT connectivity in remote location, wide area sensors, data concentrator of metering infrastructure require long-range communication.

It is noteworthy that all of the aforementioned protocols work in the unlicensed band and are thereby cost-effective. However, issues such as interferences due to the coexistence of signals from other technologies in the vicinity, device density, coverage, quality of service (QoS), and security must be considered.

3.2.1 State of the Art

In a smart city, the candidate communication protocols can be divided into two: (1) long-range and (2) short- and medium-range protocols. A widely applied long-range protocol is LTE/LTE-U.On the other hand, some of the many short- and medium-range protocols are WiFi, ZigBee, DSRC, and NBIoT. These protocols are further described below:

1. **LTE-U**: LTE-U or licensed assisted access of LTE is a long-range protocol which is currently used predominantly for personal mobile communication, but can be effectively employed for other long-range communication needs in a future smart city [16–18]. Smart devices such as smart phones, laptops, tablets, smart health monitoring sensor (i.e., fitbit) along diverse sensor networks of the cities are connected to Internet through LTE. Due to already established infrastructure, LTE can support communication for a wide range. Usually, LTE utilizes licensed spectrum for guaranteed quality of service. On the other hand, LTE-U operates in the unlicensed bands, coexisting with other protocols such as WiFi and ZigBee. How-

ever, a proper spectrum sharing technique is required for fair neighbor-sharing of spectrum.

2. **WiFi**: WiFi is the most popular short-range communication protocol standard, which has manifested as wireless local area network (WLAN) for smart phones, laptops, personal computers, smart TVs, cameras, and much more [19]. WiFi is based on the IEEE Standard 802.11 and has a range of $20m$ under indoor scenarios. Modern WiFi systems can yield a throughput of up to $1GB/s$. WiFi operates in unlicensed bands like $2.4GHz$ and $5GHz$. However, for long range, node density, node velocity, interference, and performance degraded. Since WiFi provides Internet and IoT connectivity for smart devices, it is very critical for vibrant socioeconomic climates for the citizens of smart cities.

3. **ZigBee**: ZigBee is the protocol that is most feasible to meet low power and data rate requirements of smart city [16]. Various devices such as smart home appliances, smart meters, and sensors can utilize ZigBee for short-range communication. Specially, advanced metering infrastructure of smart cities utilizes ZigBee to relay consumption data from meter-to-meter and finally to data concentrator. The maximum throughput of ZigBee is $250Kb/s$, and its range is $90m$.

4. **DSRC**: DSRC is a medium-range protocol designed for V2V communication. In this case, the communication latency should be around $1ms$, whereas reliability should be as high as 99.99%. It uses a $5.9GHz$ band with a bandwidth of $75MHz$. The typical range of DSRC is $100m$. It can ensure autonomous driving, collision avoidance, proper platooning of vehicles, optimization of traffic signal etc.

5. **NBIoT**: NBIoT is a low-power wide area network (WAN) designed for a large number of devices distributed across a geographical location [14, 20, 21]. It uses the communication resources of LTE, builds an IoT network for its supporting devices, and can use the dedicated/unused resources of LTE.
Since NBIoT can support long-range communication and does not require to connect base station each time similar to LTE, it can provide communication network for sensor nodes requiring seamless and long-range communication.

3.2.2 Open Research Challenges

Although diverse communication protocols are proposed for various applications in smart cities, there are some challenges which need to be addressed:

1. **Spectrum**: Most of the short-range communication protocols operate in the unlicensed bands—900 MHz, 2.4 GHz, 3.5 GHz, and 5 GHz. The fact that these bands are free for any application to use is a critical issue here [22–24]. This is so because all protocols like WiFi, ZigBee, DSRC, and LTE-U use the same band and might operate simultaneously, causing interference to each other's transmission and reducing quality.

2. **Coverage**: If the nodes/devices are located over a larger territory, coverage can become a issue. Short-range communication can support up to a range of $100m$

above which, data must be relayed using repeaters and amplifiers. While it might appear to be a viable solution, other issues such as propagation delay and corruption of packets while relaying become prominent. On the other hand, use of long-range communication could increase costs of installation and maintenance.

3. **Node density**: In smart cities, diverse sensors along with various applications can pose a burning issues. For example, if a heavy vehicle crosses a bridge, all sensors on the bridge can report data about the vehicle simultaneously, which can jam a poorly designed network. Hence, density-aware network design for IoT is one of the still open research challenges that must be done considering worst-case scenarios.

4. **Heterogeneous Network**: In unlicensed bands, diverse technologies will coexist for operation which is a big challenge. For instance, in the case of coexisted operation of LTE-U and WiFi, proper band sharing mechanism needs to be proposed [18], which if done improperly will cause one protocol to degrade the performance of the other.

5. **Security**: Security is also a major issue that hampers effective smart city communication. The real-time AMI smart meter data can reveal sensitive details about the behavior and lifestyle patterns of customers at homes, which if compromised would pose critical threats to personal safety [25, 26]. Therefore, resilient communication alone is not enough, but added security mechanism to it must be ensured to truly realize the vision of smart city. Therefore, resilient communication alone is not enough, but added security mechanism to it must be ensured to truly realize the vision of smart city.

3.3 Decentralized Data Processing

The smart city is a complex, interconnection of heterogeneous networks which generate data continuously as streams or periodic bursts. This data, when accumulated over time, presents significant challenges with respect to processing, storage, and maintenance. Data in general has five major characteristics and are more crucial for a smart city because of their direct implications to its resilience and security. *Volume* refers to the size of the generated data that typically grows to the orders of terabytes and petabytes quickly. *Velocity* is a measure of data resolution and varies between devices. For example, synchrophasors like PMUs generate phasor data between 10 and 120 samples per second while smart meters generally report data every fifteen minutes. EVs, on the other hand, might transmit data in short bursts to update the regional traffic management system. *Variety* refers to the heterogeneity of data sources while *veracity* corresponds to the accuracy and integrity of the data measured. Hence, veracity can be considered a measure of data quality. Finally, data *validity* is the measure of its freshness. Considering the smart city CI networks are dynamic and evolving in nature, their response to system events and changes must also be dynamic, especially because a data point generated a few milliseconds ago can become outdated and useless if not used in a timely manner. This is true, especially for real-time applications such as traffic control, fault identification, switching operations, or absorbing and injecting reactive power to balance load and demand. Such latency-cognizant applications need to have constant and on-demand access to data.

3.3.1 State of the Art

In the traditional sense, cloud infrastructure has empowered users and applications, with access to data each is entitled to, through customized online services at any place, anytime. Albeit centralized in architecture, it brings different users their data, applications, and services under one shed where sharing and resourcefulness are leveraged to everyone's advantage. This has made the cloud a lucrative solution for revolutionizing the way data is ingested, processed, stored, and maintained. For example, cloud-based energy system solutions have been extensively studied in the literature. However, significant challenges with this framework like privacy of data, communication latency and congestion, and resource contention still persist. To address these concerns, Cisco, and then the OpenFog Consortium of which Cisco is a member, introduced, defined, and presented the requirements for a fog computing approach which is decentralized. Defined formally, it is "a horizontal, system-level architecture that distributes computing, storage, control, and networking functions closer to the users along a cloud-to-thing continuum" [27]. It can be inferred from this definition that the fog is an optimal use of both cloud as well as edge analytics, as illustrated by Fig. 3.2. It shifts simple, lightweight, low-latency applications to the

Fig. 3.2 Decentralized fog computing for future smart city analytics

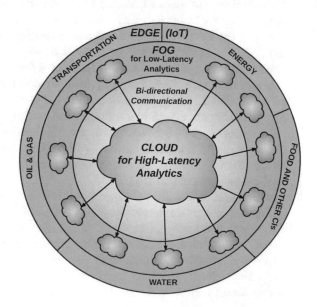

edge but still recommends complex, heavyweight, high-latency applications to be located on cloud. This way, the intelligence is distributed but not entirely decentralized to the point where end devices are burdened with computation-intensive jobs. This also helps making an effective use of resources at the cloud, reduces congestion in the communication links, and introduces a hierarchical approach to processing and analyzing smart city big data.

3.3.2   Open Research Areas

The fog architecture can be described by nine key interdependent characteristics which have also become the major avenues of emerging research [28, 29].

1. **Reliability**: Application of dynamic resource allocation and deallocation, parallel computing, and virtualization to ensure network availability, coverage, and IoT sensor service uptime.
2. **Hierarchical structure**: A layered approach to processing data with descriptive analytics done at the edge using fog nodes, and more complex predictive and/or prescriptive analytics done at the cloud.
3. **Resilience**: Fog nodes by design are dynamic and interoperable with peer-to-peer communications that enable another fog node to takeover a compromised, failed, or overloaded fog node.
4. **Dynamism and heterogeneity**: The dynamic resource allocation and deallocation capability of fog nodes make them a viable option to meet the ever-changing customer needs within a smart city.
5. **Partitioning**: The fog architecture is cloud-inclusive, and hence, applications must be aware of what data pertains to the fog nodes and what to the cloud. For example, applications that need real-time, short-term data must be moved to the fog nodes while those that need aggregated data that spans over a longer time period for offline processes must be retained at the cloud.
6. **Interoperability, automation, and autonomy**: Fog supports interoperability at all levels (device to application), enables autonomy of resources, and services for discovery and registration, management, security, and operation. It also provides looser coupling between interdependent functions to ensure scalability, parallelization, and functional autonomy through the use of software-defined networks (SDNs), message queuing telemetry transport (MQTT) and constrained application protocol (CoAP) [30].
7. **Latency and bandwidth**: Considering this was the impetus for the conception of fog, the devices are inherently energy conservative, resource-constrained, and bandwidth-aware, and enforce validity and mission criticality.
8. **Security and Privacy**: Fog nodes uphold the complex balance between ensuring security and maintaining privacy by regulating access to information hierarchically. The nodes at lower level send only processed information to the higher level, and that too, in a secure manner. Thus, the security of data grows as it passes from

the edge to the cloud. To ensure a given fog node that can trust the data coming from a node in the same or lower layer, it must implement a Root of Trust (RoT).

9. **Agility**: In the future smart city, intelligence can only be leveraged when different infrastructure can talk to one another and understand each other's data. This is achieved through context-aware analytics. An example of context-aware analytics can be viewed for a microgrid supported by a large-scale RES installation like solar and comprising both residential as well as commercial loads and EVs. Here, a fog computing network can be made operational which takes into account data from solar inverters, microgrid status from frequency disturbance recorders (FDRs), weather forecast from meteorological station, and expected profiles of different loads to perform associative analysis between them and generate optimal mix of generation sources (probably a combination of solar, energy storage, and external grid) to meet the expected demand. Such contextual understanding expedites the ability of models to make faster and optimal decisions that then drive broader decisions and policy management higher in the hierarchy.

3.4   Smart Energy Design

A progressive drive toward smarter energy design for managing dynamic loads of a smart city is crucial. Energy for smart city dwellers is expected to be heterogeneous and diverse, compounded by the inclusion of a higher renewable generation mix. At the same time, the demands of future cities are also expected to disproportionately increase, since energy is required to drive the added intelligence to other CI networks like transportation, analytics, and communication, for instance. Due to the required maintenance of real-time situation awareness (SA), distributed sensors such as PMUs, AMI smart meters, automatic feeder switches (AFSs), and other IEDs across the grid need to have adequate computation, communication, and security resources, which require significant expense of energy.

There are also applications that leverage the internetworking between smart city networks. Some example scenarios include: real-time monitoring of EV traffic congestion patterns to allocate corresponding energy for associated charging needs, mapping the communication network contingencies and bottlenecks to efficiently reroute network traffic through idle nodes (and thus balancing the load distributed across the processing components), and offering incentives to residents such as Time of Use (ToU) electricity pricing, net metering, and direct load control. These are some smart energy designs that could be worth exploring in the future and will be a significant step forward in the envisioned goal of future smart cities poised to improve quality of life and promote a sustainable lifestyle. In this section, state of the art in smart energy design is elaborated briefly using three use-cases: (1) maintaining real-time SA of the grid to ensure its stability and reliability using PMUs, (2) guaranteeing more power and choice to end-customers through applications such as net metering, ToU pricing, community RESs, and direct load control, and (3) addressing dynamic demands with fluctuating generation through the use of secondary-generation and high-speed analytics.

3.4.1 State of the Art

3.4.1.1 Real-Time Situation Awareness

Synchrophasor devices such as PMUs, phasor data concentrators (PDCs), and phasor gateways (PGWs) are implemented, traditionally on transmission line systems and recently also on distribution feeder systems, to provide wide area monitoring, protection, and control, all of which contribute to real-time SA. A typical PMU comprises multiple modules, each dedicated to a specific purpose. The current and voltage signals from current and potential transformers, respectively, are used to measure their three-phase values, which are then converted to digital data. A microprocessor-based module compiles these values along with the time stamp, computes the phasors, and synchronizes them with Coordinated Universal Time (UTC) standard reference used by the Global Positioning System (GPS) receivers that in turn acquire a time lag based on the atomic clock of the GPS satellites. PMUs also measure the local frequency and its rate of change, and with further customization, it can measure individual phase voltage and current along with the harmonics, negative and zero sequence values. The information thus generated helps paint a dynamic picture of smart grid at any given time.

Multiple standards currently exist for data measurement, transfer, and communication among synchrophasors, proposed by IEEE and IEC [31–34]. However, due to the prevalence of multiple specifications and guidelines, the literature also hints at possible contradictions in the standard recommendations [35–39]. Specifically, IEEE Standard C37 series deals with synchrophasors, with Standard C37.118 being widely used for data measurement and transfer.

The Oak Ridge National Laboratory (ORNL) has been leading the FNET/GridEye project since 2004, as part of which FDRs have been installed and managed in many countries across the world to capture dynamic behaviors of the grid. FDRs are modified versions of traditional PMUs which are connected at 110 V and hence incur lower installation costs [40]. It is noteworthy that FDRs are largely deployed at grid-tied RES installations, are capable of measuring $1,440$ samples per second, and use GPS receivers that have an accuracy of up to $1\,\mu s$ for synchronization. FDRs deployed as part of the FNET/GridEye project use the Internet to send data directly to the data servers at University of Tennessee and ORNL for analytics.

If there are multiple PMUs within a confined area such as substation, local PDCs are installed to aggregate site-level PMU data and then transmit them to the Super-PDC located generally at the control center. IEEE Standard C37.244–2013 provides PDC requirements. While PDCs were traditionally used only to aggregate PMU data and monitor the constituent WAMS, increasing scale and volume of data needs has made PDCs heavier on data processing and storage as well [41]. PDCs are software applications that conduct various data quality checks and set necessary flags according to the issues encountered, perform self-performance logging and alarming, data validation, transformation, scaling, normalization, protocol conversion, and much more. PDCs usually have an associated host computer which helps with local processing, descriptive analytics, localized visualization, and storage of PMU data. There is typically a direct interface between the PDC and the utility supervisory control And data acquisition (SCADA) or energy management system (EMS).

3.4.1.2 More Power to Customers

One of the most powerful business models is the one that places the customer at its center. This model executes processes and enforces policies around customer interests and offers incentives, choice, transparency, privacy, and security. A well-evolved smart city can be thought to employ such a model for its daily operations. Since its networks are interdependent, exchange of information becomes easier, and interoperability at the levels of devices, functions, services, and architecture becomes the norm. This will enable the prosumers to have greater access to important information such as energy consumption patterns, weather forecasts, neighborhood traffic profiles, and future energy needs. Some key customer-centric services today are net metering, Time of Use electricity pricing, community renewable generation, and direct load control [42, 43].

In scenarios where a prosumer generates more energy than they need from RES installations, **net metering** allows them to sell the surplus to the utility by directing it back to the grid, thereby offsetting the price they pay for the electricity they purchase at another time within the same billing cycle. The meters used in such systems can record bidirectional power flow, thereby measuring only the "net" energy consumed by the users that the utility would bill them for. This means the net energy can go negative in case the prosumer's net generation surpassed net consumption. In such cases, the customers become eligible for energy credits where the residual offset from one billing cycle gets carried over to the next. However, the concept of net metering is highly driven by policies, which vary among countries, and sometimes even across the states of a country.

While net metering bills the customers based on the *net amount* of electricity that was consumed or produced, **ToU pricing** is based on *when or at what specific time* the electricity was consumed or produced. While some utilities fix ToU prices on a seasonal basis, others use daily demand profiles. A utility might charge higher for electricity usage during peak daytime loads and lower during minimum daytime loads. Typically, the price of electricity falls during nighttime when the grid is not strained. When ToU pricing and net metering are combined, however, the customer is usually at a loss. For example, the surplus energy generated might allowed to be sold to the grid only at a price that is equal to the ToU price at that time. It implies that the customer cannot make profitable money from selling the surplus energy as the net energy never goes negative over the billing cycle. While in case of purely net metered systems, generation exceeding consumption implies net negative power, in a net metered-ToU system, generation exceeding consumption can bring the net power at most to zero, thus providing the customer with no credits.

Community RESs and **direct load control** will be major ways in which supply and demand can be shared among a larger, connected community of users in a city. In a closely knit community of users, each with rooftop solar and/or additional renewable sources of generation, surplus power can be sold to other customers within the same neighborhood instead of to the utility. Sharing renewable generation within a community can reduce the net demand to be serviced by a utility's feeder and relieve its strains during peak daytime load hours. Alternatively, the utility can undertake proactive measures to manage the loads of customers that have agreed to participate in their direct load control program. In this scenario, the customer allows the utility to shut off specific loads in their households during specific periods of time to minimize the overall strain on the servicing feeder. For example, the utility could shut off demanding appliances such as dishwashers and EV charging stations during peak load hours of an evening and reschedule them for later at night when the load is considerably less. The utility might choose to use the cutoff load for serving other critical customers such as hospitals and industries. It is also a win–win for the customers since rescheduling certain applications to low-load hours implies lower charges on electricity usage. However, this has also recently raised privacy concerns, considering the utility will have a direct control over the customer usage patterns and preferences.

3.4.1.3 Addressing Dynamic Demands

In the future smart city, energy will not only be required to power homes and industries, but also to drive other CI networks like transportation, communication infrastructure, and data analytic engines at the command and control centers. As identified in Fig. 3.1, this complex interdependency between energy, transportation, analytics, communication, and other networks is an evident requirement that also poses a significant challenge for implementation. With the use of distributed intelligence, power requirements for devices on the grid's edge also increase. This is so because they now have to support preliminary to moderate capabilities like data crunching, cleaning, formatting and structuring, filtering, quality checking, and statistical analyses. Ubiquitous, bidirectional peer-to-peer communication between distributed devices has lately elevated the significance of service availability over security.

For example, in a scenario where the utility dynamically manages the city's demands through an approach that combines the principles of net metering, ToU pricing, direct load control, and sharing RES generation among customers within a community, there is an overarching requirement that the sensing, monitoring, protection, and control devices of the corresponding networks be reliable, available, fault-tolerant, and robust. In order to achieve this, the devices themselves must be adequately powered and equipped with backups depending on their criticality to the intended mission. Hence, this use-case can be thought of as an umbrella that covers the previous two use-cases as well and brings out the interdependency between energy and other networks of a smart city.

3.4.2   Open Research Challenges

In this subsection, open research challenges pertinent to smart energy design in the context of future cities are briefly discussed.

1. **Lack of standardization**: While there exist many standards for interconnecting, operating, protecting, controlling, and maintaining core power systems, the same cannot be said about how those power systems interact with computation, communication, and other CI networks of smart cities. An example would be the lack of accepted standards for smart grid big data management, which is an area only recently being looked into by key players like IEEE, NIST, and ISO/IEC. Another aspect of standardization is vendor agnosticism where the developed solutions and methodologies can be uniformly applied across different interdependent networks without significant alterations to their underlying mechanisms.
2. **Optimality of computations**: Most of the power system applications are driven by one or more objectives which are contingent on system and resource constraints. These optimization algorithms have to be accurate as well as quick, considering both are stringent constraints for power system scenarios. While both heuristic as well as deterministic models have been vastly applied to solve different power system problems, the former suffers from poor convergence accuracy while the latter takes a long time to converge. Solutions to problems such as unit commitment and optimal power flow, demand forecasting, weather estimation, and load shedding which exhibit elevated levels of optimality and convergence speeds are now a hotly researched area.
3. **Cyber-physical security**: Due to the ubiquity and pervasiveness of analytics and communication, IoT devices, especially of the energy network, have an increased attack surface and have lately been successfully targeted by many attackers [44]. The recent cyber-attack at the power grids of Ukraine had immense impacts on the physical realm of it, causing cascading failures across its cities [45]. Despite an abundance of vulnerability detection tools, frequent patching and automatic reconfiguration services, attacks continue to evolve and sometimes outwit the defense mechanisms. Lack of SA has been identified by regulatory bodies like the North American Electric Reliability Commission (NERC) as one of the key causes for the high success rates of attacks, but the underlying issue is lack of *efficient* SA, one where operator requirements are correctly captured and efficient analytics are in place to leverage useful information from the wealth of data and present them to the operators in a timely manner.
4. **Scalability**: The smart grid is one of the largest smart city networks, and solutions to address its challenges are required to be scalable. These solutions must be

capable of handling data of different sizes, types, and speeds. In the long run, they must adapt to population growth, urbanization, changes in culture, politics and demographics, and must be adaptive and agile to meet both individuals as well as community goals.

5. **Privacy concerns**: As important as security is the privacy of customer data, which can be fostered only through transparency, mutual trust, and assurance of information integrity. It is anticipated that smart cities of the future will evolve from placing their trust on central authorities into decentralized, peer-to-peer trust models. To this effect, blockchains have been identified as a possible candidate for ensuring trust and maintaining integrity and privacy of different smart city network devices. This is, however, a nascent area of study with untapped potential for further research and development.

6. **Social aspect**: Although efficient SA can help improve the responsiveness of defenders toward mitigation of attacks, the overall effort could still be futile if the operators are not cognitively at par with the information being displayed to them. It is compounded by the fact that the responsiveness of humans to system events is much slower than that by automated tools. However, ironically, irrespective of the countless automated tools in a cybersecurity pipeline, the decision making is always the responsibility of a human operator. This necessity and the gap therein has been highlighted by NERC in its Critical Infrastructure Protection (CIP) guidelines, where a special emphasis is given to human behavior and the impact of different visualization methods on decision making [46, 47].

7. **Policymaking and enforcement**: The greatest challenge to the mass deployment and application of any smart city technology boils down to the Government's policymaking and the effectiveness of its enforcement. During policymaking stages, it is important to involve communities and grassroots organizations that reflect the real needs of customers. Comprehensive surveys, interviews, and discussions must be held to cover citizens of all demographics, value their interests, negotiate their differences, and understand their key requirements.

3.5   Smart Transportation

Over the past century, there have been dramatic developments in the electric vehicle (EV) industry. EVs are considered as the key elements in the electrification of transportation systems which can significantly reduce carbon footprints of the energy consumed in transportation systems. Electrified transportation systems can lead to a cleaner and smarter transportation system and therefore are of great interest for future smart cities. EVs are considered as mobile sources of energy which can establish bidirectional power connections with other infrastructures including power grid, residential and commercial buildings. Specifically, grid-to-vehicle (G2V) and vehicle-to-grid (V2G) connections can enhance the resiliency and stability of the power system. In extreme conditions, vehicle-to-infrastructure (V2I), vehicle-to-vehicle (V2V), and vehicle-to-home (V2H) connections can be established to enhance the power resiliency of a the smart city. For example, in extreme weather conditions with widespread power outages, EVs can be used as power sources for a local critical infrastructure such as communication infrastructures, airports, and hospitals.

3.5.1 State of the Art

Inductive electric vehicle (EV) charging based on inductive power transfer (IPT) is an emerging technology that can revolutionize the transportation systems. This technology provides more convenience to the use of EVs by eliminating the need for interaction of the driver in the charging process of EVs. Since electrical connections between the vehicle and the AC mains are eliminated, inductive EV charging systems are not affected by rain, snow, dust, and dirt and therefore are considered as a safer alternatives for conductive charging systems. This technology can be mainly categorized into two main types: dynamic (in-motion) IPT and static IPT. Static inductive EV charging enables automated charging process without the need for human interaction. A typical static inductive charging system is presented in Fig. 3.3. Such systems can be used in any residential and commercial parking station including house garages, stores. This type of inductive charging also can be implemented on the roadway behind the traffic lights where vehicles have minimum movement for a short time, providing extra range for on-road EVs.

The requirements for light-duty static inductive EV charging systems are established in SAE TIR J2954 Standard [48]. This standard defines four inductive charging levels as 3.7, 7.7, 11, and 22kW for light-duty EVs with a minimum power transfer efficiency of 85%. Also, the standard requires the inductive charging system to

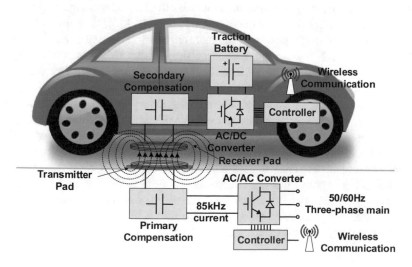

Fig. 3.3 Typical structure of a static EV charging system

Fig. 3.4 Typical structure of a roadway dynamic EV charging system

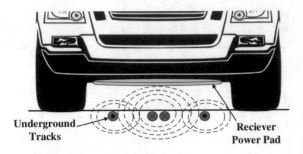

Underground Tracks

Reciever Power Pad

have a single nominal operating frequency of 85 kHz. Although the requirements for static inductive charging for heavy-duty EVs are not defined in the standard, it is successfully implemented in many studies [49–51].

On the other hand, dynamic IPT systems are designed to establish a contactless power transfer between EVs and the power grid. This technology can be implemented in roadways to significantly increase the range of EVs by providing continues inductive power while the vehicle is moving. Using dynamic IPT systems, the size of the EV traction batteries which are the most expensive parts in EVs can be reduced and thereby the production costs of EVs are decreased. A typical structure of a dynamic EV charging system is shown in Fig. 3.4. The system is composed of multi-conductor underground transmitter tracks, a receiving power pad mounted under the EV. In order to achieve a convenient dynamic EV charging, the system should be able to tolerate high levels of EV misalignments respect to the underground tracks [52, 53].

Various magnetic structures and topologies are presented. These structures are usually composed of ferrite core material, shielding plate, Litz wire coil, and coil former. The structures can be categorized into two main types: unipolar and bipolar magnetic structures. Unipolar structures include circular and rectangular power pads [54–57]. Bipolar magnetic structures include solenoid [58], double-D (DD) [59], double-D-Q (DDQ) [60] types. Unipolar pads have simpler structures and are mainly used in static inductive charging systems. Using bipolar magnetic structures, high misalignment tolerance can be achieved and therefore they are preferred choice in dynamic inductive charging systems.

Any inductive charging system should be designed to have maximum power transfer efficiency, and its electromagnetic field (EMF) emissions should comply with EMF exposure standards as defined by International Commission on Non-Ionizing Radiation Protection (ICNIRP) [61].

Similar to conductive charging systems, inductive EV charging systems are capable of bidirectional power transfer which enables V2G, G2V, V2I, and V2V connections through the IPT system. Since such inductive connections can be easily and quickly established at different geographical locations of the power system, they can make the power system more complex and dynamically varying.

3.5.2   Allocation of Electric Vehicle Charging Stations in Smart Cities

Large-scale utilization of EVs in smart cities brings about challenges to the power systems by causing power quality and reliability issues. Specifically, unplanned EV charging processes will result in many issues in distribution power systems such as reliability issues and power loss increase. These issues can be mitigated by optimal power system planning to overcome high demands of EVs in power delivery systems. In [62], a new method for finding the optimal simultaneous allocation of EV charging stations and renewable energy sources (RES) and handing EV charging processes is presented. The optimal solutions are found by formulating and solving a multi-objective optimization problem (MOP) that considers different objectives such as demand supply cost efficiency, EV battery maintenance costs, power losses, and voltage fluctuations. The proposed MOP is solved using a hybrid Genetic Algorithm-Particle Swarm Optimization (GA-PSO) to obtain the optimal allocation of charging stations in a case study power system. It is shown that the optimal allocation of EV charging stations significantly enhances the power quality, power losses, and voltage profiles.

The optimal allocation of EV charging stations based on reliability indices of power system is presented in [63]. This is achieved by the use of a probabilistic model for the distribution of EVs in a smart city. A composite reliability index which is a linear combination of different power system reliability indices is used as the main objective function. The performed analysis shows that the optimal location of EV charging stations highly depends on the total number of EVs and the EV charging method. The results also show that the optimal allocation of charging stations can significantly improve the reliability of the power system.

3.5.3   Open Research Challenges

Interoperability: The interoperability between EVs and different infrastructures of a smart city is crucial for achieving a resilient, reliable system. This requires establishing a set of interoperability standards for G2V, V2G, V2I, V2H, and V2V connections. It is important to note that these connections can be either a conductive or an inductive connection. The requirements and alignment methodology for inductive static EV charging systems are presented in SAE TIR J2954. However, the standard only describes G2V connections while inductive V2G connections still remain an open research area.

Communication: Both static and dynamic inductive charging/discharging systems require a two-way communication between EV and infrastructures to exchange the information such as EV ID, state of charge (SOC) of the EV battery, charging/discharging authorization, billing information, command and control signals. The location and point of connection of the EV is of great importance as it can be used to determine the requirements of the connection. For example, to establish a connection between power grid and an EV, the type of the connection (G2V/V2G) and the charging/discharging power can be determined based on the geographical location of the EV and grid requirements.

Foreign Object Detection: IPT systems can induce high levels of power into metallic objects which bring about fire hazard. Also, living objects if entered to the charging area can be exposed to high levels of EMF emissions. Therefore, inductive charging systems should be equipped with a foreign object detection (FOD) mechanism that can detect the presence of metallic and living objects in the charging area and shut down the charging process if necessary.

Implementation cost of Dynamic IPT system: The main barrier in the development of dynamic inductive EV charging system in roadways is the implementation cost. Such systems should be highly optimized for power transfer efficiency, cost efficiency, and vehicle misalignment tolerance with EMF emission considerations [57].

Economic Allocation of EV Charging Stations: The problem of optimal allocation of EV charging stations by considering the economic aspect of power system is open area that requires further analysis and investigation [63]. This can be achieved by adding cost efficiency of the system as another objective to the EV charging station multi-objective optimization problem.

3.6   Summary and Conclusion

Smart cities of the future will integrate different networks in a synergistic manner and manage them jointly to achieve harmonious broader goals. While the research on these networks has been isolated thus far, the interdependency between them can no longer be ignored. Factors such as high penetration of RESs, the interconnectivity between food, energy, and water, interplay between energy, oil, and gas, and the underlying backbone of communication and data analytics, form the various drivers to adopt an interdependent, cross-functional approach to conducting research in this area. The chapter highlighted the following findings in its comprehensive survey.

- In a smart city, its constituent networks engage in complex interactions and information exchange over a robust communication infrastructure augmented by high-end data processing, optimization, control and visualization models, thus making it a system of interdependent networks
- To ensure the sustainability of these networks, they must be constantly optimized, secured, made reliable, and flexible to the changing needs and demographics of customers
- Five important communication protocols that will potentially find great applications in a smart city environment are LTE-U for long-range communication, WiFi

for short-range communication, ZigBee for low-power communication, NBIoT for low-power communication between large-scale distributed IoT network devices, and DSRC for medium-range communication between EVs within a fleet

- These protocols use the same unlicensed bands, raising concerns such as interference due to the co-existence of signals from other technologies in the vicinity, device density, coverage, QoS, and security
- It is important for a communication infrastructure to be resilient and flexible to change, but it is more important for it to be secure. In the context of decentralized computing and rising scale of smart city networks, the communication channels and underlying protocols must become more robust against attacks, espionage, and thefts of sensitive information.
- Future smart cities will significantly shift computing intelligence from the cloud-driven centralized architecture to fog-driven distributed architecture
- Smart city big data attributes are not just limited to velocity, volume, and variety, but also include veracity and validity, owing to the system being dynamic and continuously evolving. The developed solutions must take these two attributes into concern too
- A hierarchical need-driven data transmission approach is considered feasible for a smart city environment, where edge devices pass only those data points that the devices higher in the hierarchy would need. Implementing security at each level strengthens the overall information and improves its quality
- Context-aware data analytics expedites the ability of optimization and control models to make faster decisions that then drive broader decisions and policymaking higher in the hierarchy
- Information exchange is expected to be point-to-point, which raises concerns of interoperability, standardization, security, and reliability
- Three of the most fundamental and indispensable use-cases to the energy network of future smart cities were considered: maintaining real-time SA using synchrophasor devices, guaranteeing more power and choice to end-customers, and addressing dynamic demands using secondary-generation and high-speed optimization
- Prosumers in the future will share an equal responsibility of producing services too, becoming prosumers. Smart city solutions of the future need to be prosumer-driven
- Net metering, ToU dynamic electricity pricing, direct load control, and community RESs are some of the prosumer-driven scenarios of a smart city energy network
- Lack of standardization, ensuring global optimum as well as speed of convergence in power system optimizations, delivering on human-in-the-loop cybersecurity, and effectively inculcating social aspects to energy delivery are some of the key open research challenges identified
- The integration EVs with critical infrastructures of a smart city can be realized with various connections including G2V, V2G, V2I, V2H, and V2V connections
- Inductive charging/discharging technology based on IPT systems is an emerging technology that can revolutionize the smart transportation systems in smart cities

- Dynamic inductive charging systems which can be implemented in roadways can increase the range of the EVs and are considered as potential solution for mass electrification of transportations systems

To establish and sustain smart cities by the year 2030, a holistic vision to effectively integrate these interdependent networks is required. The chapter addresses this challenge by introducing and describing intelligent communication, decentralized analytics, smart energy design, and smart transportation for future city scenarios before addressing the key open research challenges in each of these areas. By doing this, the chapter aims to serve as a starting point for researchers entering the domain of smart city and are interested in conducting cross-functional research across its interdependent networks covered by the chapter.

Acknowledgements The authors would like to acknowledge that the work for this chapter is supported by the following grants: CNS-1553494 (NSF) and 800006104 (DOE).

References

1. DOE, The smart grid: An introduction. A U.S. Department of Energy (DOE) Technical Report (2009)
2. A. Anzalchi, A. Sarwat, A survey on security assessment of metering infrastructure in smart grid systems. SoutheastCon **2015**, 1–4 (2015)
3. S. M. Amin, Electricity infrastructure security: toward reliable, resilient and secure cyber-physical power and energy systems, in *IEEE PES General Meeting*, pp. 1–5, July 2010
4. L. Wei, A.H. Moghadasi, A Sundararajan, A.I. Sarwat, Defending mechanisms for protecting power systems against intelligent attacks, in *System of Systems Engineering Conference (SoSE), 2015 10th*, pp. 12–17, May 2015
5. A. Sanjab, W. Saad, I. Guvenc, A. Sarwat, S. Biswas, Smart grid security: threats, challenges, and solutions. [cs.IT], June 2016. arXiv:1606.06992
6. A. Sundararajan, A. Pons, A.I. Sarwat, A generic framework for eeg-based biometric authentication, in *12th International Conference on Information Technology-New Generations*, pp. 139–144, Apr 2015
7. B. McMillin, Complexities of information security in cyber-physical power systems, in *IEEE/PES Power Systems Conference and Exposition*, Mar 2009
8. K. Booroojeni, M.H. Amini, A. Nejadpak, T. Dragicevic, S.S. Iyengar, F. Blaabjerg, A novel cloud-based platform for implementation of oblivious power routing for clusters of microgrids. IEEE Access 607–619 (2016)
9. K.G. Booroojeni, M.H. Amini, S.S. Iyengar, Smart Grids: Security and Privacy Issues (Springer International Publishing, 2017)
10. A.I. Sarwat, M.H. Amini, A. Domijan, A. Damnjanovic, F. Kaleem, Weather-based interruption prediction in the smart grid utilizing chronological data. J. Mod. Power Syst. Clean Energy 607–619 (2015)
11. A.I. Sarwat, A. Domijan, M.H. Amini, A. Damnjanovic, A. Moghadasi, Smart grid reliability assessment utilizing boolean driven markov process and variable weather conditions, in *North American Power Symposium*, Oct 2015
12. J. Mandic-Lukic, B. Milinkovic, N. Simic, Communication solutions for smart grids, smart cities and smart buildings, in *Mediterranean Conference on Power Generation, Transmission, Distribution and Energy Conversion (MedPower 2016)*, pp. 1–7 (2016)

13. J. Jin, J. Gubbi, S. Marusic, M. Palaniswami, An information framework for creating a smart city through internet of things. IEEE Internet Things J. **1**, 112–121 (2014)
14. I. Parvez, M. Jamei, A. Sundararajan, A.I. Sarwat, Rss based loop-free compass routing protocol for data communication in advanced metering infrastructure (ami) of smart grid, in *2014 IEEE Symposium on Computational Intelligence Applications in Smart Grid (CIASG)* (IEEE, 2014) pp. 1–6
15. I. Parvez, *Data Transmission in Quantized Consensus*. Ph.D. Thesis, University of North Texas, 2013
16. I. Parvez, N. Islam, N. Rupasinghe, A.I. Sarwat, Gven, Laa-based lte and zigbee coexistence for unlicensed-band smart grid communications, in *SoutheastCon 2016*, pp. 1–6 (2016)
17. M. Sriyananda, I. Parvez, I. Güvenç, M. Bennis, A.I. Sarwat, Multi-armed bandit for lte-u and wifi coexistence in unlicensed bands, in *2016 IEEE Wireless Communications and Networking Conference (WCNC)* (IEEE, 2016) pp. 1–6
18. I. Parvez, M. Sriyananda, İ. Güvenç, M. Bennis, A. Sarwat, Cbrs spectrum sharing between lte-u and wifi: a multiarmed bandit approach. Mob. Inf. Syst. **2016** (2016)
19. I. Parvez, T. Khan, A. Sarwat, Laa-lte and wifi based smart grid metering infrastructure in 3.5 ghz band, in *IEEE R10HTC 2017*
20. R. Ratasuk, B. Vejlgaard, N. Mangalvedhe, A. Ghosh, Nb-iot system for m2m communication, in *2016 IEEE Wireless Communications and Networking Conference*, pp. 1–5, Apr 2016
21. I. Parvez, N. Chotikorn, A.I. Sarwat, Average quantized consensus building by gossip algorithm using 16 bit quantization and efficient data transfer method, in *International conference on Intelligent Systems, Data Mining and Information Technology (ICIDIT)*, pp. 1–5
22. S. Bhattarai, J.M.J. Park, B. Gao, K. Bian, W. Lehr, An overview of dynamic spectrum sharing: Ongoing initiatives, challenges, and a roadmap for future research, *IEEE Transactions on Cognitive Communications and Networking*, June 2016
23. M.G.S. Sriyananda, I. Parvez, I. Güvene, M. Bennis, A.I. Sarwat, Multi-armed bandit for lte-u and wifi coexistence in unlicensed bands, in *2016 IEEE Wireless Communications and Networking Conference*, pp. 1–6 (2016)
24. I. Parvez, A. Sarwat, Frequency band for han and nan communication in smart grid, in *2014 IEEE Symposium on Computational Intelligence Applications in Smart Grid (CIASG)* (IEEE, 2014)
25. I. Parvez, A. Islam, F. Kaleem, A key management-based two-level encryption method for ami, in *2014 IEEE PES General Meeting—Conference & Exposition*, pp. 1–5 (IEEE, 2014)
26. I. Parvez, A.I. Sarwat, L. Wei, A. Sundararajan, Securing metering infrastructure of smart grid: a machine learning and localization based key management approach. Energies **9**(9), 691 (2016)
27. OpenFog, Openfog reference architecture for fog computing, *OpenFog Consortium Architecture Working Group* (2017)
28. F.Y. Okay, S. Ozdemir, A fog computing based smart grid model, in *In Proceedings of the International Symposium on Networks, Computers and Communications (ISNCC16)* (2016)
29. C.C. Byers, P. Wetterwald, Fog computing distributing data and intelligence for resiliency and scale necessary for iot the internet of things, in *ACM Symposium on Ubiquity* (2015)
30. D. Locke, Mq telemetry transport (mqtt) v3. 1 protocol specification. IBM developerWorks Tech. Libr. (2010)
31. IEEE, Ieee standard for intelligent electronic devices cyber security capabilities. IEEE Power Energy Soc. (2013)
32. IEEE, IEEE standard cybersecurity requirements for substation automation, protection, and control systems. IEEE Power Energy Soc. (2014)
33. IEEE, IEEE trial-use standard for a cryptographic protocol for cyber security for substation serial links. IEEE Power Energy Soc. (2011)
34. S.R. Firouzi, H. Hooshyar, F. Mahmood, L. Vanfretti, An iec 61850-90-5 gateway for ieee c37.118.2 synchrophasor data transfer, *NASPI-ISGAN International Synchrophasor Symposium* (2016)

35. H. Mehta, Will the current set of standards assure measurement subsystem interoperability? in *NASPI PSTT Meeting* (2011)
36. A. Johnson, Standards associated with synchrophasors, in *IEEE PES General Meeting* (IEEE)
37. K. Martin, Synchrophasor standards: support and development, in *DOE/OE Transmission Reliability Program* (IEEE, 2015)
38. I. Ali, M.A. Aftab, S.M.S. Hussain, Performance comparison of iec 61850-90-5 and IEEE c37.118.2 based wide area pmu communication networks. J. Mod. Power Syst. Clean Energy (MPCE) (2016)
39. H. Retty, Evaluation and standardizing of phasor data concentrators, Apr 2013
40. J. Chai, Y. Liu, J. Guo, L. Wu, D. Zhou, W. Yao, Y. Liu, T. King, J.R. Gracia, M. Patel, Wide-area measurement data analytics using fnet/grideye: a review, in *Power Systems Computation Conference* (IEEE, 2016)
41. IEEE, Ieee guide for phasor data concentrator requirements for power system protection, control, and monitoring. IEEE Power Energy Soc. (2013)
42. A. Poullikkas, G. Kourtis, I. Hadjipaschalis, A review of net metering mechanism for electricity renewable energy sources. Int. J. Energy Environ. 975–1002 (2013)
43. A. Gholian, H. Mohsenian-Rad, Y. Hua, Optimal industrial load control in smart grid. IEEE Trans. Smart Grid 2305–2316 (2015)
44. NIST, National institute of standards and technology framework for improving critical infrastructure cybersecurity. *NIST*, Jan 2017
45. T. Rueters, Cyberattack that crippled ukrainian power grid was highly coordinated. CBC News **11** (2016)
46. NERC, North american electric reliability corporation critical infrastructure protection compliance standards, in *NERC* (2017)
47. FERC, "Federal energy regulatory commission cyber & grid security," *FERC*, 2005
48. J. Schneider, Sae tir j2954 wireless charging of electric and plug-in hybrid vehicles. SAE Int
49. J.M. Miller, A. Daga, Elements of wireless power transfer essential to high power charging of heavy duty vehicles. IEEE Trans. Transp. Electr. **1**, 26–39 (2015)
50. R. Bosshard, J.W. Kolar, Multi-objective optimization of 50 kw/85 khz ipt system for public transport. IEEE J. Emerg. Sel. Top. Power Electron. **4**, 1370–1382 (2016)
51. J.H. Kim, B.S. Lee, J.H. Lee, S.H. Lee, C.B. Park, S.M. Jung, S.G. Lee, K.P. Yi, J. Baek, Development of 1-mw inductive power transfer system for a high-speed train. IEEE Trans. Ind. Electron. **62**, 6242–6250 (2015)
52. G.A. Covic, J.T. Boys, Modern trends in inductive power transfer for transportation applications. IEEE J. Emerg. Sel. Top. Power Electron. **1**, 28–41 (2013)
53. C.C. Mi, G. Buja, S.Y. Choi, C.T. Rim, Modern advances in wireless power transfer systems for roadway powered electric vehicles. IEEE Trans. Ind. Electron. **63**, 6533–6545 (2016)
54. M. Budhia, G.A. Covic, J.T. Boys, Design and optimization of circular magnetic structures for lumped inductive power transfer systems. IEEE Trans. Power Electron. **26**, 3096–3108 (2011)
55. R. Bosshard, U. Iruretagoyena, J.W. Kolar, Comprehensive evaluation of rectangular and double-d coil geometry for 50 kw/85 khz ipt system. IEEE J. Emerg. Sel. Top. Power Electron. **4**, 1406–1415 (2016)
56. M. Moghaddami, A. Anzalchi, A. Moghadasi, A. Sarwat, Pareto optimization of circular power pads for contactless electric vehicle battery charger, in *2016 IEEE Industry Applications Society Annual Meeting*, pp. 1–6, Oct 2016
57. M. Moghaddami, A. Anzalchi, A.I. Sarwat, Finite element based design optimization of magnetic structures for roadway inductive power transfer systems, in *2016 IEEE Transportation Electrification Conference and Expo (ITEC)*, pp. 1–6, June 2016
58. M. Chigira, Y. Nagatsuka, Y. Kaneko, S. Abe, T. Yasuda, A. Suzuki, Small-size light-weight transformer with new core structure for contactless electric vehicle power transfer system, in *2011 IEEE Energy Conversion Congress and Exposition*, pp. 260–266, Sept 2011
59. M. Budhia, J.T. Boys, G.A. Covic, C.Y. Huang, Development of a single-sided flux magnetic coupler for electric vehicle ipt charging systems. IEEE Trans. Ind. Electron. **60**, 318–328 (2013)

60. G.R. Nagendra, G.A. Covic, J.T. Boys, Sizing of inductive power pads for dynamic charging of evs on ipt highways. IEEE Trans. Transp. Electr. **3**, 405–417 (2017)
61. J. Lin, R. Saunders, K. Schulmeister, P. Söderberg, A. Swerdlow, M. Taki, B. Veyret, G. Ziegelberger, M.H. Repacholi, R. Matthes et al., Icnirp guidelines for limiting exposure to time-varying electric and magnetic fields (1 hz to 100 khz). Health Phys. **99**, 818–836 (2010)
62. M.R. Mozafar, M.H. Moradi, M.H. Amini, A simultaneous approach for optimal allocation of renewable energy sources and electric vehicle charging stations in smart grids based on improved ga-pso algorithm. Sustain. Cities Soc. **32**, 627–637 (2017)
63. M.H. Amini, A. Islam, Allocation of electric vehicles' parking lots in distribution network. ISGT **2014**, 1–5 (2014)

Chapter 4
Interdependent Interaction of Occupational Burnout and Organizational Commitments: Case Study of Academic Institutions Located in Guangxi Province, China

Xiazi Sun

4.1 Background

Nowadays, instructors in colleges play a more and more important role gradually whereas they are encircled in a dilemma. It is reported instructors in colleges perceive high level of stress, high workload, role conflict and ambiguity, low predictability, lack of participation and social support, and experienced unfairness. Most of instructors deem this job merely as a transition for a further career. On the other hand, loss of passion and patience is quite common which is adverse to individuals themselves nor students in educational area. Therefore, we choose this group as participants for burnout research. Occupational burnout refers to a persistent negative state of mind resulting from prolonged work stress, referring to a report from China Development of Human Resource 2005. The level of burnout of professionals in education is the third top in ranking (which indicates the public officers and employees in logistics are the top two occupations successively in this ranking).

Meanwhile, this report from China Development of Human Resource 2005 illustrates more than 85% instructors' tenure of job is less than 5 years. For organizations, it possibly indicates that instructors are unstable group to educational construction and it is at high wastage rate of human resource.[1]

---

[1]Zhu Hong. Causes and Countermeasures of Job Burnout On instructors in colleges. Hebei University, 5, 2010.

X. Sun (✉)
SYSU-CMU Shunde International Joint Research Institute,
Eastern Nanguo Road, Shunde 528399, Guangdong, China
e-mail: sunxiazi@sdjri.com; 2evenxiazi@sina.com
URL: http://www.sdjri.com/English/

© Springer International Publishing AG, part of Springer Nature 2018
M. H. Amini et al. (eds.), *Sustainable Interdependent Networks*,
Studies in Systems, Decision and Control 145,
https://doi.org/10.1007/978-3-319-74412-4_4

In Guangxi Province (in southwestern China), recent years, the growth of employment of instructors is far slower than the increase of students' admission. According to the ordinances of National Ministry of Education, the ratio of population of instructors to the population of students is suggested not to be less than 1:35. But in fact the ratio is only 1:200 in Guangxi,[2] as the expansion of faculty position hardly grows while universities enroll students fast-increasingly. As a result, a lot of stressors generate such high level of burnout among instructors in colleges in Guangxi.

Hence, this research targets the group of instructors in colleges. It is related study on the relation of occupational burnout and organizational commitments. Hopefully, it takes advantage to seek solutions of occupational hazard both for individual himself and for organizational management.

4.2 Introduction of Definitions [1–75]

4.2.1 Instructors in College

Worldwide, the job of instructors in colleges is likely a characteristic occupation as the staff in colleges that is only existed in China. Its responsibility is intervenient by professors and administrative staff. Instructors are in charge of students for their both study and daily life. This position plays an extremely pivotal role by supervision in communicating students and schools (organization).

4.2.2 Burnout

Many studies that have focused on stress have also focused on the concept of burnout. A review of the literature on this construct has yielded consistent agreement about the effects of this condition, although inconsistent agreement about the nature of its cause. In this summary, the concept of burnout is not discussed as it is defined consistently as follows.

Maslach (1982) has identified and defined burnout as a "syndrome of emotional exhaustion, depersonalization or dehumanization, and diminished personal accomplishment that can occur among individuals who do people work of some kind." Meanwhile, the three factors encompassing the construct of burnout by Maslach and Jackson (1981): emotional exhaustion (feelings of exhaustion and

---

[2]Liang Pihuan. The Structural dilemma of Occupational Burnout in Colleges in Guangxi and Coping Strategies. The Journal of The Communist Party School in Zhengzhou, 6 (186–188), 2008.

being emotionally overextended with work), depersonalization (unfeeling or impersonal reactions toward patients or clients), and diminished personal accomplishment (feelings of incompetence in work) are determined.[3]

4.2.3 Organizational Commitments

Organizational commitments refer to an individual's overall feelings about the organization. They are the psychological bond with an organization that an employee has, and they have been found to be related to behavioral investments in the organization, likelihood to stay with the organization, and goal and value congruence (Mowday, Steers, and Porter, 1982).

Meyer and Allen (1991) have termed the three factors' component of organizational commitments: normative commitment (a strong belief in and acceptance of the organizational commitment), affective commitment (a willingness to exert considerable effort on behalf of the organization), and continuance commitment (a strong desire to maintain membership in the organization).[4]

Likewise, the concept is not discussed in the summary. However, organizational commitments are defined by which are component of five factors adjusted and adapted to China social background by Prof. Lin Wenshuan (凌文辁)(2001). The level of organizational commitments is measured with normative commitment, affective commitment, aspiring commitment, economic commitment, and opportunistic commitment.[5]

4.3 Hypotheses

Based on the participants who are instructors in colleges (universities) of Guangxi Province:

Hypothesis 1: The level of burnout among individuals differed via distinctions of gender, age, years of working, marital status, and academic category will be discrepant significantly.
Hypothesis 2: The degree of organizational commitments among individuals differed via distinctions of gender, age, years of working, marital status, and academic category will be discrepant significantly.

---

[3]Maslach, C.A Multidimensional Theory of Burnout. Cooper C L. Theory of Organizational Stress. London: OxfordUniversity Press, 2001:68–85.

[4]Wang Yanfang. Related Study of Occupational Burnout and Organizational Commitments Among Teachers in Kindergarten. Educational Development Research, 2(70–74), 2007.

[5]Bian Ran. Relation of Occupational Burnout, Organizational Commitments and Quit Intention. Huazhong Normal University, 2004.

Hypothesis 3: Occupational burnout will be negatively associated with organizational commitments.
Hypothesis 4: Occupational burnout will be predictable significantly to organizational commitments.

4.4 Method

4.4.1 Participants and Procedure

Questionnaire paperwork was taken to the school units among eight universities (basically there are in total six national universities in Guangxi and two more key colleges included in the investigation), distributed to instructors by researcher or the supervisors in the units. Once the questionnaires were completed in half or one hour, paperwork was returned directly to the researcher that was considered a high rate of collection.

Of the eligible participants (n = 130) who returned the questionnaire (n = 126, response rate 96.9%), 43 participants are male and 83 are female, 92 are under 30 years old and 34 are older than 30 years old, 78 have worked on the same position under 3 years and 27 for 3–5 years and 21 for more than 5 years, 45 are unmarried and 79 are married and 2 is divorced. In the same time, data are classified into three academic categories of art, technological (engineering), and medical school which the participants are employed in.

4.4.2 Measure

Burnout was measured basically by the Maslach Burnout Inventory-Education Survey Model (MBI-ES). Occupational commitment questionnaire is referred to Modway, Steers, and Porter's OCQ series. Both of the questionnaires are adapted more pertinent to the regional research (Table 4.1).

MBI-ES: The items were scored on a 7-point frequency rating scale ranging from 0 (never) to 6 (daily). High scores for exhaustion and cynicism and low scores for diminished personal accomplishment are indicative of burnout. The items of diminished personal accomplishment were reversed (diminished professional efficacy).

Table 4.1 Reliability coefficient of questionnaires

Type	Cronbach's alpha	Items	N
MBI-ES	0.696	17	126
OCQ	0.676	23	126

Table 4.2 Descriptive statistics of burnout among instructors

	N	Min	Max	Mean	SD	V
EE	126	5.00	30.00	18.381	5.395	29.102
DP	126	4.00	20.00	10.151	4.188	17.537
DPA	126	12.00	30.00	18.841	4.142	17.159
Sum	126	22.00	75.00	47.373	10.206	104.156

EE emotional exhaustion, *DP* depersonalization, *DPA* diminished personal accomplishment

OCQ: The items were scored on a 5-point frequency rating scale ranging from 0 (totally disagree) to 5 (totally agree). High scores for five factors such as normative commitment, affective commitment, aspiring commitment, economic commitment, and opportunistic commitment are indicative of high organizational commitment degree (Table 4.2).

4.4.3 Data Analysis

4.4.3.1 Occupational Burnout

(1) General characteristic,

(2) The demographic variables for each factor of burnout included gender (Table 4.3), age (Table 4.4), years of working (Table 4.5), marital status (Table 4.6), academic category (Table 4.7).

Table 4.3 Regression analysis of burnout via gender variables

	Sex	N	Mean	SD	SE Mean	t	Sig
EE	M	43	16.791	5.180	0.790	−2.428	0.898
	F	83	19.205	5.348	0.587	−2.453	
DP	M	43	9.186	4.777	0.729	−1.880	0.007
	F	83	10.651	3.782	0.415	−1.747	
DPA	M	43	19.209	3.204	0.489	0.716	0.003
	F	83	18.651	4.560	0.500	0.799	
Sum	M	43	45.186	10.898	1.662	−1.745	0.779
	F	83	48.506	9.702	1.065	−1.682	

EE emotional exhaustion, *DP* depersonalization, *DPA* diminished personal accomplishment

Table 4.4 Regression analysis of burnout via age variables

	Age	N	Mean	SD	t	Sig
EE	<=30	92	17.750	5.3298	−2.005	0.601
	>30	33	19.909	5.252	−2.018	
DP	<=30	92	9.815	3.628	−1.329	0.000
	>30	33	10.939	5.420	−1.106	
DPA	<=30	92	19.076	4.251	1.047	0.380
	>30	33	18.206	3.820	1.100	
Sum	<=30	92	46.641	9.275	−1.328	0.078
	>30	33	49.353	12.319	−1.167	

EE emotional exhaustion, *DP* depersonalization, *DPA* diminished personal accomplishment

Table 4.5 Regression analysis of burnout via years of working variables

	Years	N	Mean	SD	F	Sig
EE	1–3	78	18.423	5.720	0.085	0.918
	3–5	27	18.037	6.009		
	5–10	21	18.667	2.955		
DP	1–3	78	10.231	3.924	0.686	0.505
	3–5	27	10.630	5.562		
	5–10	21	9.238	2.982		
DPA	1–3	78	19.449	4.353	2.454	0.090
	3–5	27	18.185	4.086		
	5–10	21	17.429	2.908		
Sum	1–3	78	48.103	9.871	0.650	0.524
	3–5	27	46.852	13.370		
	5–10	21	45.333	6.053		

EE emotional exhaustion, *DP* depersonalization, *DPA* diminished personal accomplishment

Table 4.6 Regression analysis of burnout via marital status variables

	Status	N	Mean	SD	F	Sig
EE	U	45	18.356	5.568	0.013	0.987
	M	79	18.380	5.398		
	D	2	19.000	0.000		
DP	U	45	9.733	3.447	1.017	0.365
	M	79	10.468	4.577		
	D	2	7.000	0.000		
DPA	U	45	19.556	3.481	1.169	0.314
	M	79	18.481	4.489		
	D	2	17.000	0.000		
Sum	U	45	47.644	7.088	0.198	0.821
	M	79	47.329	11.749		
	D	2	43.000	0.000		

U unmarried, *M* married, *D* divorced
EE emotional exhaustion, *DP* depersonalization, *DPA* diminished personal accomplishment

Table 4.7 Regression analysis of burnout via academic variables

	Subject	N	Mean	SD	F	Sig
EE	Art	67	16.761	5.030	11.670	0.000
	Tech	21	17.714	6.092		
	Medical	38	21.605	4.175		
DP	Art	67	9.597	3.730	2.566	0.081
	Tech	21	9.619	5.352		
	Medical	38	11.421	4.071		
DPA	Art	67	19.269	4.467	1.979	0.143
	Tech	21	17.238	4.024		
	Medical	38	18.974	3.437		
Sum	Art	67	45.627	9.919	6.145	0.003
	Tech	21	44.571	10.962		
	Medical	38	52.000	8.914		

EE emotional exhaustion, *DP* depersonalization, *DPA* diminished personal accomplishment

Table 4.8 Descriptive statistics of organizational commitment among instructors

	N	Min	Max	Mean	SD	V
ASC	126	8.00	22.00	14.714	2.635	6.942
EC	126	8.00	19.00	14.524	2.369	5.611
NC	126	7.00	20.00	14.429	2.847	8.103
OC	126	6.00	19.00	10.818	2.673	7.142
AFC	126	11.00	22.00	17.706	2.404	5.777
Sum	126	45.00	89.00	72.191	7.143	51.019

ASC aspiring commitment, *EC* economic commitment, *NC* normative commitment, *OC* opportunistic commitment, *AFC* affective commitment

4.4.3.2 Organizational Commitment

(1) General characteristic (Table 4.8).

(2) The demographic variables for each factor of organizational commitment included gender (Table 4.9), age (Table 4.10), years of working (Table 4.11), marital status (Table 4.12), and academic category (Table 4.13).

4.4.3.3 Association between occupational burnout and organizational commitment was analyzed referred to Table 4.14

Table 4.9 Regression analysis of organizational commitment via gender variables

	Sex	N	Mean	SD	SE Mean	t	Sig.
ASC	M	43	15.465	1.279	0.195	2.343	0.000
	F	83	14.325	3.049	0.335	2.943	
EC	M	43	14.163	2.449	0.373	−1.234	0.823
	F	83	14.711	2.319	0.255	−1.213	
NC	M	43	14.837	2.672	0.407	1.161	0.425
	F	83	14.217	2.926	0.321	1.196	
OC	M	43	10.070	1.857	0.283	−2.299	0.003
	F	83	11.205	2.946	0.323	−2.641	
AFC	M	43	17.558	2.481	0.378	−0.497	0.669
	F	83	17.783	2.374	0.261	−0.490	
Sum	M	43	72.093	6.206	0.946	−0.110	0.712
	F	83	72.241	7.618	0.836	−0.117	

ASC aspiring commitment, *EC* economic commitment, *NC* normative commitment, *OC* opportunistic commitment, *AFC* affective commitment

Table 4.10 Regression analysis of organizational commitment via age variables

	Age	N	Mean	SD	SE Mean	t	Sig.
ASC	<=30	92	14.446	2.429	0.253	−1.902	0.182
	>30	34	15.441	3.047	0.523	−1.714	
EC	<=30	92	14.457	2.322	0.242	−0.523	0.348
	>30	34	14.706	2.517	0.432	−0.504	
NC	<=30	92	14.750	2.780	0.290	2.114	0.566
	>30	34	13.559	2.884	0.495	2.078	
OC	<=30	92	10.315	2.555	0.266	−3.636	0.081
	>30	34	12.177	2.540	0.436	−3.645	
AFC	<=30	92	17.739	2.158	0.225	0.251	0.164
	>30	34	17.618	3.005	0.515	0.216	
Sum	<=30	92	71.707	7.061	0.736	−1.254	0.507
	>30	34	73.500	7.304	1.253	−1.234	

ASC aspiring commitment, *EC* economic commitment, *NC* normative commitment, *OC* opportunistic commitment, *AFC* affective commitment

4.4.3.4   Regression analysis was adjusted stepwise for the three factors of burnout as independent variables while the five factors of organizational commitment (aspiring commitment, economic commitment, normative commitment, opportunistic commitment, and affective commitment) as dependent variables as referred to Tables 4.15, 4.16, 4.17, 4.18, and 4.19

Table 4.11 Regression analysis of organizational commitment via years of working variables

	Years	N	Mean	SD	F	Sig
ASC	1–3	78	14.244	2.445	4.359	0.015
	3–5	27	15.037	3.180		
	5–10	21	16.048	2.085		
EC	1–3	78	14.333	2.410	0.812	0.446
	3–5	27	15.000	2.869		
	5–10	21	14.619	1.244		
NC	1–3	78	14.372	2.820	1.943	0.148
	3–5	27	15.222	3.215		
	5–10	21	13.619	2.247		
OC	1–3	78	10.449	2.676	2.026	0.136
	3–5	27	11.519	3.274		
	5–10	21	11.286	1.309		
AFC	1–3	78	17.308	1.797	2.954	0.056
	3–5	27	18.259	3.778		
	5–10	21	18.476	1.861		
Sum	1–3	78	70.705	7.071	4.818	0.010
	3–5	27	75.037	8.258		
	5–10	21	74.048	3.81413		

ASC aspiring commitment, *EC* economic commitment, *NC* normative commitment, *OC* opportunistic commitment, *AFC* affective commitment

4.4.4 Interview

Instructors (n = 18, response rate 14.3%) selected randomly from participants (n = 126) were interviewed. The outline of talk mainly aimed on the internal motivation in the job, the attitude and suggestions to the organization, the stressors sensed, the intention, and the expectations of one's own career.

4.5 Discussion

4.5.1 Occupational Burnout

Generally, the result indicates mild occupational burnout (mean = 47.373, Table 4.2) among instructors in Guangxi colleges, not as severe as it was predicted. Additionally, severe burnout is relatively more common among women, individuals over the age of thirty, novice instructors working within 3 years, unmarried individuals, and those employed in art schools, compared to the rest. In separate dimensional analyses, the different level of depersonalization and diminished personal accomplishment between male and female reveals significant (Table 4.3).

Table 4.12 Regression analysis of organizational commitment via marital status variables

	Status	N	Mean	SD	F	Sig
ASC	U	45	14.533	2.341	1.682	0.190
	M	79	14.734	2.781		
	D	2	18.000	0.000		
EC	U	45	14.044	2.576	3.446	0.035
	M	79	14.709	2.185		
	D	2	18.000	0.000		
NC	U	45	14.644	2.479	3.086	0.049
	M	79	14.190	2.983		
	D	2	19.000	0.000		
OC	U	45	10.089	2.712	3.160	0.046
	M	79	11.177	2.596		
	D	2	13.000	0.000		
AFC	U	45	18.133	1.392	3.447	0.035
	M	79	17.380	2.770		
	D	2	21.000	0.000		
Sum	U	45	71.444	6.989	6.272	0.003
	M	79	72.190	6.830		
	D	2	89.000	0.000		

U unmarried, *M* married, *D* divorced
ASC aspiring commitment, *EC* economic commitment, *NC* normative commitment, *OC* opportunistic commitment, *AFC* affective commitment

Another significant association relating to the dimensionality of depersonalization reveals participants above 30 years old experience severe burnout than younger participants (Table 4.4, sig < 0.000). Meanwhile, the individuals work in art schools experience more emotional exhaustion than the other participants (Table 4.7, sig < 0.000). Remarkably, the instructors employed at medical school show the highest severe burnout (Table 4.7).

Not surprisingly the result indicates women experience severe burnout. It may be explained by at least two issues. Firstly, normally women might be due to balance between family and occupation. Most of them have to input partial time and energy in family. Yet the workload at school is the same as male workmates. Secondly, according to the interview, female instructors' achievement motivation relates lower and less expectation on this job. It supports the result of the lower level of personal accomplishment among women, whereas it interprets less diminished personal accomplishment than men which means mild burnout.

In China, the instructor in college is one of the commendable positions for graduates. The job assists the students who are young group indeed; hence, it is not deemed to be a long-term job for majority when the employees are growing older. This is the consideration that for participants the ages are merely gapped into two items in the study. Basically, most instructors can get promotion after being an

Table 4.13 Regression analysis of organizational commitment via academic variables

	Subject	N	Mean	SD	F	Sig
ASC	Art	67	14.313	2.548	1.904	0.153
	Tech	21	15.476	1.436		
	Medical	38	15.000	3.171		
EC	Art	67	14.493	2.092	3.401	0.037
	Tech	21	15.619	3.090		
	Medical	38	13.974	2.236		
NC	Art	67	14.463	2.836	0.480	0.620
	Tech	21	13.905	1.446		
	Medical	38	14.658	3.419		
OC	Art	67	11.030	2.132	0.471	0.626
	Tech	21	10.476	3.386		
	Medical	38	10.632	3.106		
AFC	Art	67	17.463	2.265	1.780	0.173
	Tech	21	17.381	2.801		
	Medical	38	18.316	2.361		
Sum	Art	67	71.761	6.597	0.266	0.767
	Tech	21	72.857	7.696		
	Medical	38	72.579	7.873		

ASC aspiring commitment, *EC* economic commitment, *NC* normative commitment, *OC* opportunistic commitment, *AFC* affective commitment

Table 4.14 Result of association analysis

	ASC	EC	NC	OC	AFC	Sum
EE	−0.053	0.170	−0.191*	0.407***	−0.167	0.057
DP	−0.335***	0.094	−0.409***	0.200*	−0.445***	−.330***
DPA	−0.219*	−0.136	−0.285***	0.088	−0.393***	−0.339***
Sum	−0.254**	0.073	−0.384***	0.333***	−0.430***	−0.243**

* = $P<0.05$, ** = $P<0.01$, *** = $P<0.001$)

Table 4.15 ASC-burnout regression analysis result

	B	t	R	R^2	$\triangle R^2$	F	Sig.
SEM			0.335[a]	0.112	0.105	15.640	0.000[a]
Con	16.852	28.841					0.000
DP	−0.211	−3.955					0.000

ASC aspiring commitment, *DP* depersonalization

Table 4.16 EC-burnout regression analysis result

	B	t	R	R^2	$\triangle R^2$	F	Sig.
SEM			0.234[a]	0.055	0.031	2.355	0.000[a]
Con	14.820	12.499					0.000
DPA	−0.161	−2.433					0.016

EC economic commitment, *DPA* diminished personal accomplishment

Table 4.17 NC-burnout regression analysis result

	B	t	R	R^2	$\triangle R^2$	F	Sig.
SEM			0.409[a]	0.167	0.160	24.811	0.000[a]
Con	17.249	28.214					0.000
DP	−0.278	−4.988					0.000

NC normative commitment, *DP* depersonalization

Table 4.18 OC-burnout regression analysis result

	B	t	R	R^2	$\triangle R^2$	F	Sig.
SEM			0.407[a]	0.166	0.159	24.641	0.000[a]
Con	7.110	9.137					0.000
EE	0.202	4.964					0.000

OC opportunistic commitment, *EE* emotional exhaustion

Table 4.19 AFC-burnout regression analysis result

	B	t	R	R^2	$\triangle R^2$	F	Sig.
SEM			0.445[a]	0.198	0.192	30.691	0.000[a]
Con	20.301	40.087					0.000
DP	−0.256	−5.540					0.000

AFC affective commitment, *DP* depersonalization

instructor for few years. Thus, those who have not been assigned to a new position might sense crisis on career. Burnout which is common and severe on elder instructors over thirty years old can be interpreted on the dimensionality of emotional exhaustion and depersonalization. Because of the diverse constituents of the job, performance of instructors is assessed ambiguously sometimes. It causes widely diminished personal accomplishment but not significant difference among the age variables.

Concerning the medical expertise and professional, emotion might be tense for both students and staff at medical schools. The instructors get high scores on the dimensionality of emotional exhaustion and depersonalization. High pressure is the primary stressor related to medical academy. Furthermore, the instructors in technological school associated more diminished personal accomplishment. Which has been noticed, referring to the interview, most instructors in technological school graduated from technological background as well, but their work currently mainly focuses on management and administration. It causes anxiety and alienation on their professionals, which is hardly for them to get accomplishment from job.

4.5.2   Organizational Commitment

In recapitulation, the score of organizational commitments among instructors in Guangxi colleges demonstrates an above average degree (Table 4.8, mean = 72.191, SD = 7.143). Emotional commitment shows comparatively high degree on top, while opportunistic commitment stands on the bottom. High degree of organizational commitments is relatively more common among women, individuals over the age of thirty, instructors working within 3–5 years, and those employed in technological schools, compared to the rest. Though Table 4.12 indicates divorced instructors got higher score of organizational commitments, considering there are only two complied divorced participants, the validity would be discussed.

In separated dimensional analyses, the difference of aspiring commitment and opportunistic commitment between male and female reveals significant (Table 4.9). Male instructors pay more attention to personal accomplishment in work, higher aspiring commitment attributes to achieving expectation. According to the interview, more male participants regard this job as an occupational transition to further career, compared to female instructors. Women tend to rest on this job steadily, which supports the result of higher opportunistic commitment in organization.

Particularly, instructors under 30 years old show low opportunistic commitment. This group presents greater occupational indeterminacy. Participants working as instructors over 10 years perceive higher opportunistic commitment based on reliance interest on organization.

4.5.3   Association Analysis

Referring to Table 4.14, general occupational burnout negatively related to general organizational commitments (Correlation = −0.243, p < 0.01), readily interpreting high occupational burnout causes low organizational commitments.

4.6   Conclusion

Instructors in colleges (universities) of Guangxi Province:

Conclusion 1: The level of occupational burnout among individuals differed via distinctions of gender, age, and academic category is discrepant significantly, while it is not discrepant significantly among instructors via distinctions of years of working, and marital status. Hypothesis 1 validates partially.

Conclusion 2: The degree of organizational commitments among individuals differed via distinctions of gender, years of working, marital status, and academic category is discrepant significantly. Ages classified do not indicate significantly as an active factor on discrepancy. Hypothesis 2 validates partially.

Conclusion 3: Occupational burnout confirms negatively associated with organizational commitments.

Conclusion 4: The level of occupational burnout can predict the degree of organizational commitments effectively.

4.7 Limitations of This Study

In terms of methodology and statistics, this study has several potential limitations. First, though the data collection is based on several main colleges and universities, anyhow it was not composed by all high education institutes where supply instructor positions in Guangxi Province. Secondly, domestically, Guangxi is an comparatively underdeveloped area, and its diverse culture and historical background probably could help to explain the result of analysis. Thus, the survey is apt to be regional research hardly to generalize to all area. Third, it was considered on relation between only two items and there is the possibility of simplifying employ–organization interactions from the actuality.

References

1. T. Barrier, R.C. King, Examination of the correlates of burnout in information systems professionals. Inf. Resour. Manag. J. **12**(3), 5–13 (1999)
2. P.L. Brill, The need for an operational definition of burnout. Fam. Community Health **6**, 12–24 (1984)
3. B. Buchanan, Building organizational commitment, the socialization of managers in work organization. Adm. Sci. Q. **19**, 533–546 (1974)
4. J.P. Burton, T.W. Lee, B.C. Hotom, The influence of motivation to attend, ability to attend and organizational commitment on different types of absence behaviors. J. Manag. Issues **11**(2), 181–197 (2002)
5. B. Byrne, Investigating causal links to burnout for elementary, intermediate and secondary teachers, in *Paper presented at the Annual Meeting of the American Educational Research Association, San Francisco, CA*, vol. 4 (pp. 20–24, 1992)
6. C. Cherniss, *Professional Burnout in Human Service Organizations* (Praeger, New York, 1980)
7. C.L. Cordes, T.W. Dougherty, A review and an integration of research on job burnout. Acad. Manag. Rev. **18**(4), 621–656 (1993)
8. I.L. Densten, Re-thinking burnout. J. Organ. Behav. **22**(81), 833–847 (2001)

9. D. Etzion, Burnout: The Hidden Agenda of Human Distress (No.IIBR Series in Organizational. Behavior and Human Resources, Working Paper No.930/87), 1987
10. R.T. Golembiewski, R.F. Munzenrider, *Phase of Burnout: Development in Concepts and Applications* (Praeger, New York, 1988)
11. L.G. Herbiniak, J.A. Alutto, Personal and role related factor, in the development of organizational commitments. Adm. Sci. Q. **17**, 560–599 (1972)
12. R. King, V. Sethi, The moderating effect of organizational commitment on burnout in information system professionals. Eur. J. Inf. Syst. **6**, 86–96 (1997)
13. R.T. Lee, C.M. Maslach, The impact of interpersonal environment on burnout and organization commitments. J. Organ. Behav. **9**, 297–308 (1988)
14. J.E. Mahtieu, D.M. Zajac, A review and meta-analysis of the antecedents, correlates, and consequences of organizational commitments. Psychol. Bull. **108**(2), 171–194 (1990)
15. C. Maslach, M.P. Leiter, *The Truth about Burnout* (Jossey-Bass, San Francisco, 1997), pp. 1–14
16. C. Maslach, S.E. Jackson, The measurement of experienced burnout. J. Occupational Behavior **2**, 99–113 (1981)
17. C. Maslach, M.P. Leiter, *The Truth about Burnout: How Organizations Cause Personal Stress and What To Do About It* (Jossey-Bass, San Francisco, California, 1997)
18. J.P. Meyer, N.J. Allen, A three-component conceptualization of organizational commitments. Hum. Resour. Manag. Rev. **1**(1), 61–89 (1991)
19. A. Pines, E. Aronson, Career burnout causes and cure (New York, London, Free Press, MacMillan, 1998)
20. L.W. Porter, F.J. Smith, The etiology of organizational commitment: a longitudinal study of initial stages of employee-organization's relationships, unpublished manuscript (1970)
21. Q. Chen, Occupational burnout research and outlook. Chin. J. Psychol. Health **10** (2004)
22. J. Lin, Summary of the study on occupational burnout theory. Orient. Corp. Cult. **5** (2010)
23. Y. Li, The latest progress and prospect of occupational burnout research. Management **4** (2009)
24. H. Chen, Causes and countermeasures of job burnout
25. W. Wei, Occupational burnout research model overview. Technol. Start Mon. **4** (2011)
26. Z. Wei, The role conflict and burnout countermeasures of college instructors, Shanghai: East China Normal University Institute of Education (2007)
27. G. Lu, College faculty organizational commitments research. Henan University Education Research Institute (2005)
28. W. Zhang, The Research on College Consulting Instructors Professionalism (Henan University Press, Kaifeng, 2007 Version)
29. MengZhaolan. General Psychology (Peking University Press, 2000 Edition)
30. Document of "Provisions on the Construction of Instructors in Ordinary Colleges and Universities" (Ministry of Education No. 24)
31. Y. Chen, QiuHaifeng. The burnout level of college instructors and elimination. J. Sichuan Univer. Sci. Technol. (Nat. Sci. Edn.)
32. D. Gong, Study on the current situation of college faculty's occupational burnout. J. Jixi Univ. **4** (2006)
33. J. Li, Enlightenment of China from the theory of abroad faculty occupational burnout. Educ. Sci. **2** (2003)
34. W. Daixian, College instructors' occupational burnout factors analysis and countermeasures. Suzhou Acad. Newsp. **4** (2006)
35. F. He, College students instructors occupational burnout analysis. J. Xinjiang Normal Univer. **6** (2007)
36. Y. Li, Occupational burnout questionnaire (MBX) introduction. Environ. Occup. Med. (12), 506–507 (2004)
37. L. Chaoping, The impact of distributional equity and assignment justice on occupational burnout. Psychol. Acad. J. **35**(5) (2003)
38. Y. Li, Occupational burnout and the measurement. Comput. Sci. **26**(3) (2003)

39. Y. Li, H. Meng, Occupational burnout structure research progress. Comput. Sci. **27**(2) (2004)
40. S. Liu, S. Qingsong, College instructors' personality features and occupational burnout analysis. J. Anqing Normal Coll. (9) (2008)
41. X. Wang, Y. Gan, Progress Abroad in Psychological Science Research (5) (2003)
42. L. Di, College instructors' occupational burnout status and countermeasures. Chin. J. Youth (6) (2007)
43. J. Zhang, Let instructional work be full of bright situation. China High. Educ. (3) (2006)
44. Y. Li, M. Wu, Occupational burnout structure research. Comput. Sci. **28**(2) (2005)
45. Q. Ma, College instructor occupational burnout exploration. J. Wuhan Inst. Technol. (6) (2007)
46. Y. Zhang, S. Xiong, S. Chen, Discussion on the image of college instructors. Exploration (2) (2006)
47. Z. Jiang et al., The organizational causes of occupational burnout and coping strategies. J. Yuxi Normal Coll. (7) (2009)
48. X. Xia, Constructing a platform to build a high level of instructors group. China High. Educ. (20) (2005)
49. Wangli, The coping research of occupational burnout based on the theory of organizational management. Henan Sociol. Sci. (3) (2009)
50. D. Lin, Initial research on college instructor occupational burnout. Educ. Res. (3) (2008)
51. J. Qu, Y. Wu, Focus on the construction of a professional group of instructors. Educ. Theory Route (9) (2006)
52. X. Yang, Study on current situation, problems and countermeasures of the construction of political instructors in colleges and universities. Educ. Theory Route (9) (2006)
53. Chen Daohua. College instructors' occupational burnout cause out of the counter measures. Ideol. Polit. Educ. (2) (2007)
54. H. Liu, Analysis of college instructors' occupational burnout problem and counter measure. J. Hunan Nat. Coll. (3) (2009)
55. Y. Tong, College instructors' occupational burnout phenomenon perspective: causes and interventions. J. Sichuan Coll. Educ. (6) (2007)
56. J. Li, Progress of burnout research. J. Peking Univ. (Nat. Sci. Edn.) (2) (2003)
57. H. Zhao, Y. Liu, New ideas of professional construction among college instructors. Ideol. Theor. Educ. (11) (2005)
58. Y. Feng, X. Li, Research on the causes of occupational burnout of college faculty. J. Ideol. Polit. Res. (5) (2009)
59. Y. Chen, College instructors' occupational burnout pressure analysis and mitigation method. J. Zhejiang Chin. Med. Univ. (9) (2007)
60. J. Cao, The analysis on the survey of college instructors' occupational burnout. Educ. Sci. J. Hunan Normal Univ. (7) (2007)
61. W. Wu, Z. Wang et al., Occupational burnout and psychological management of college instructors. J. Huaxi Univ. (8) (2006)
62. W. Chen et al., Analysis based on the professionalization background of college instructors under the occupational burnout [J]. J. Hebei Normal Univ. (1) (2008)
63. L. Xiaoping, Review of organizational commitment research. Psychol. Dyn. **2**, 31–37 (1999)
64. W. Zongli, Study on the relation between occupational stress and organizational commitment in Guoming Primary School. J. Nat. Educ. **15**, 16–18 (1992)
65. Y. Guo, Organizational commitment and the cultural thinking. J. Harbin Inst. Technol. (Soc. Sci. Edn.) **3**(2), 51–58 (2001)
66. Liu Xiaopin, Wang Chongming, Organizational commitment and formation based on the different background between chinese and western cultures. Oversea Econ. Manag. **24**(1), 17–21 (2002)
67. K. Cai, Relation among organizational commitments, job satisfaction and quit intention. China Manag. Rev. **3**, 12–16 (2000)
68. Liu Xiaoping, Wang Chongming, Organizational commitment and the formation process research. Nankai Manag. Rev. **6**, 58–60 (2006)

69. Lin Wenshuan, Zhang Zhican, Fang Liluo, The causes of organizational commitment influenced. J. Psychol. **33**(3), 259–263 (2001)
70. Lin Wenshuan, Zhang Zhican, Fang Liluo, Structural model research on chinese employees. J. Manag. Sci. **3**(2), 76–79 (2000)
71. Chen Bihui, The impact of organizational commitment and occupational input on the behavior of technical staff. J. Appl. Psychol. **7**(3), 33–37 (2001)
72. W. Ling, Z. Zhang, L. Fang, Chinese workers organizational commitment research. Chin. J. Soc. Sci. **2**, 95–100 (2001)
73. Shi Zhenlei, Staff's pressure and the adjustment. J. Lanzhou Univ. **4**, 16–19 (1997)
74. Cui Xun, The impact of personal characteristics on organizational commitment and quit intention. Nankai Manag. Rev. **4**, 4–10 (2003)
75. W. Chen, H. Zhu, Study on theoccupational commitment of young faculty in colleges and universities. J. Zhejiang Educ. Inst. **5**, 91–97 (2003)

Part II
Solutions to Performance and Security Challenges of Developing Interdependent Networks

Chapter 5
High Performance and Scalable Graph Computation on GPUs

Farzad Khorasani

5.1 Introduction

Graphs provide methodical approaches to represent relationships between abstract entities and therefore are one of the most popular data representation approaches in communities dealing with massive amount of data. The relationship between people in social networks, products in a retail Web site, or roads between cities can all be expressed using graph vertices connected to one another via directed or non-directed edges. Meanwhile, processing a large graph requires a great deal of computation power. On the other hand, in the past decade GPUs have emerged as general purpose accelerators and greatly enhanced the speed and the power-efficiency of the computation in fields that expose massive amounts of data parallelism, such as scientific simulation and deep learning. Hence, the existence of parallel computation in graph analytic algorithms signals an opportunity to utilize GPUs to accelerate the process. However, efficient utilization of GPU compute power to process real-world graphs is a challenging task. Such graphs, that are extracted from real-world origins, usually exhibit a degree distribution known as power law meaning a small portion of graph vertices are connected to many other vertices while a large number of vertices have much small number of neighbors. In contrast, symmetric architecture of GPU demands data placement and access in memory in a consecutive manner. GPUs, due to their single instruction multiple threads (SIMTs) architecture, can also suffer performance loss if provided loads for its parallel components are not balanced. Therefore, the irregularity in the graph can easily lead to inefficient irregular memory accesses and load imbalance in GPUs.

In this work, we introduce some of the existing techniques that enable high-performance utilization of GPUs for processing large real-world graphs. First, in Sect. 5.2, we introduce a few graph representation strategies to store and access the

F. Khorasani (✉)
Georgia Institute of Technology, Atlanta, GA, USA
e-mail: farkhor@gatech.edu

© Springer International Publishing AG, part of Springer Nature 2018
M. H. Amini et al. (eds.), *Sustainable Interdependent Networks*,
Studies in Systems, Decision and Control 145,
https://doi.org/10.1007/978-3-319-74412-4_5

graph in GPU DRAM. Then, we introduce a dynamic task decomposition strategy for threads in the warp that enables high warp execution efficiency when processing the graph. Finally, we discuss a technique regarding scaling the graph computation to multiple GPUs.

5.2 Graph Representation Formats

In this section, we introduce a few of the well-known graph storage formats for GPUs including Compressed Sparse Row (CSR), G-Shards, and Concatenated Windows (CW) [1].

5.2.1 Compressed Sparse Row (CSR)

Representing graphs in memory to efficiently process them on GPUs has been a challenging task. Real-world graphs are usually sparse and their sizes are large involving processing over millions of vertices and edges. The latter makes it infeasible to store the graphs in space consuming representations such as the *adjacency matrix* representation.

Hence, most of existing graph processing approaches [2–4] primarily rely on the Compressed Sparse Row (CSR) format because of its compact memory footprint. For any given vertex, the CSR format allows fast access to its incoming/outgoing edges along with the source/destination vertex addresses at the other end of these edges. The representation mainly consists of four arrays:

- *VertexValues*: *VertexValues*[i] ($0 \le i < n$) represents the value of vertex v_i.
- *SrcIndxs*: *SrcIndxs*[i] ($0 \le i < m$) represents for edge e_i, the index of the source vertex in *VertexValues*. The incoming edges for a given vertex are stored in consecutive locations of this array.
- *InEdgeIdxs*: *InEdgeIdxs*[n] = m. *InEdgeIdxs*[i] ($0 \le i < n$) represents the starting index of a sub-array E_i of *SrcIndxs*. The end of this sub-array E_i can be determined by the entry at $i + 1$. E_i combined with *SrcIndxs* represents the incoming edges for node n_i.
- *EdgeValues*: *EdgeValues*[i] ($0 \le i < m$) represents the value of the edge e_i.

The neighborhood of vertex n_i can be determined by looking at locations of *VertexValues* which are represented by the sub-array starting at *SrcIndxs*[*InEdgeIndxs*[i]] and ending at *SrcIndxs*[*InEdgeIndxs*[$i + 1$]], and the edge weights can be determined by the sub-array of *EdgeValues* starting at *InEdgeIndxs*[i] and ending at *InEdgeIndxs*[$i + 1$]. Figures 5.1 and 5.2 show an example of a graph and its CSR representation using its incoming edges. As we can see, the neighborhood of vertex 2 (shown in green) is represented by *VertexValues*[1] and *VertexValues*[7] and the sub-array *EdgeValues*[4:5].

Fig. 5.1 An example graph

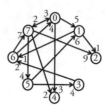

Fig. 5.2 CSR representation of graph in Fig. 5.1

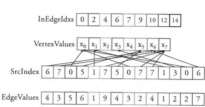

As discussed, *SrcIndxs*[*i*] and *SrcIndxs*[*i* + 1] represent non-consecutive indices in *VertexValues* array. Hence, parallel reading of these values by threads in a warp leads to random non-coalesced memory accesses requiring multiple inefficient memory transactions. Note that a major portion of graph processing is reading these vertex structures over and over again, making the problem more significant. In other words, a great number of accesses are fulfilled using greater than minimal number of transactions due to poor locality of data of interest.

5.2.2   G-Shards

Representing a graph with shards has been shown to improve disk-based graph processing performance on a shared memory system [5]. Since shards allow contiguous placement of the graph data required by a set of computations, G-Shards uses shard concept to secure benefits from coalesced accesses.

G-Shards presents a graph *G* as a set of shards where each shard is an ordered list of incoming edges and each edge *e* = (*u*, *v*) in the shard is represented by a four tuple:

- *SrcIndex*: Index of the source vertex *u*
- *SrcValue*: Content of source vertex *u*
- *EdgeValue*: Content or weight of the edge *e*
- *DestIndex*: Index of the destination vertex *v*

The set of shards used to represent *G* exhibit the following properties:

- *Partitioned*: *V* is partitioned into disjoint sets of vertices and each set is represented by a shard such that it stores all the edges having a destination is in that set.
- *Ordered*: The edges in a shard are listed based on increasing order of their *SrcIndex*.

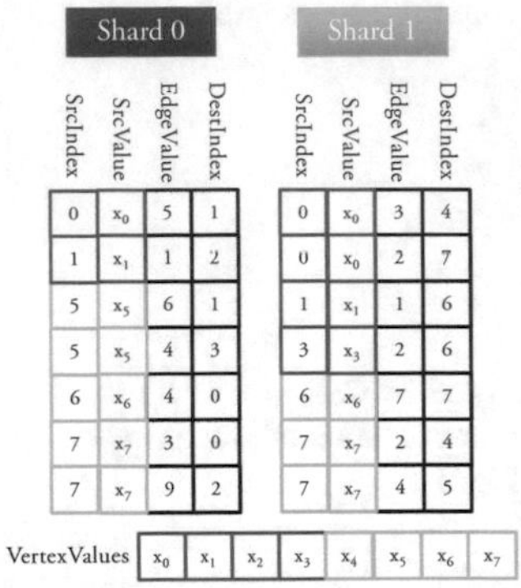

Fig. 5.3 *G-Shards*
representation of the graph
in Fig. 5.1

	Shard 0				Shard 1		
SrcIndex	SrcValue	EdgeValue	DestIndex	SrcIndex	SrcValue	EdgeValue	DestIndex
0	x_0	5	1	0	x_0	3	4
1	x_1	1	2	0	x_0	2	7
5	x_5	6	1	1	x_1	1	6
5	x_5	4	3	3	x_3	2	6
6	x_6	4	0	6	x_6	7	7
7	x_7	3	0	7	x_7	2	4
7	x_7	9	2	7	x_7	4	5

VertexValues | x_0 | x_1 | x_2 | x_3 | x_4 | x_5 | x_6 | x_7 |

Figure 5.3 shows the shard-based representation for the graph shown in Fig. 5.1. Here, the vertex set is grouped into two groups so that *Shard-0* has the list of edges whose destination is between 0 and 3 and *Shard-1* has the list of edges whose destination is between 4 and 7. Note that within *Shard-0* (and *Shard-1*), the edges are sorted based on *SrcIndex*.

To facilitate efficient processing of graphs on GPU using shards, G-Shards also maintains a separate array named *VertexValues* which allows quick access to values of vertices. Throughout the computation, *VertexValues[i]* represents the most updated value of vertex v_i.

Each shard is processed by a thread-block in the GPU in four steps. In step 1, the threads load the updated vertex values from the *VertexValues* array to the shared memory of the block. Consecutive threads of the block read consecutive elements of *VertexValues* array; hence, load requests coalesce into minimum number of transactions. In step 2, using the fetched values, the shard is processed in parallel by threads in the thread-block. Consecutive threads of the block read consecutive shard entries residing in global memory thus providing coalesced global memory loads. In step 3, the threads write back the newly computed values to the *VertexValues* array. This step is done similar to step 1 except with inversed direction of read and write. Global memory stores in this step, similar to global loads in step 1, are satisfied by minimum number of write transactions in memory controller. Step 4 (write-back stage) performs the remaining task which is to propagate computed results to other shards *SrvValue* array. In order to have coalesced global memory accesses in write-back stage as well, each warp in the block is assigned to update necessary SrvValue elements in one shard. Because of aforementioned *Ordered* property of shards, elements in one shard that need to be read and written by another shard are arranged continuously. Consecutive threads inside a warp read consecutive *SrcIndex* elements inside another shard and write to consecutive *SrcValue* elements therefore memory accesses get coalesced.

In summary, processing a shard by threads of a block on GPU involves fully coalesced global memory read and writes during all steps. Accesses to *VertexValues* in steps 1 and 3, reading shard elements in step 2, and updating necessary regions of other shards in step 4 all become coalesced.

Each shard region whose elements need to be accessed and updated together by another shard is called a computation window. A computation window W_{ij} is the set of entries in shard j that are involved during processing of shard i such that each edge in W_{ij} has *SrcIndex* in the range of vertex indices associated with shard i. This means that the source vertices of all the edges in W_{ij} belong to the vertex-range a to b if shard i represents edges whose destination vertices belong to the same range a to b. As an example, different colors for entries in Fig. 5.3 distinguish different computation windows; the windows W_{0j} are represented in red (first two elements of shard-0 and first four elements of shard-1) and windows W_{1j} are represented in green for $0 \leq j < 2$. Intuitively, for a constant k, if number of shards is p, the collection of edges in windows W_{kj}, $0 \leq j < p$, completely represents the sub-graph induced over the subset of vertices associated with shard k.

5.2.3 Concatenated Windows (CW)

G-Shards representation on GPU provides coalesced global memory accesses to neighbor contents which were not achievable in CSR format. However, the performance can be limited by various characteristics of input graphs. First, imbalanced shard sizes can cause intra-block divergence. This effect can be insignificant because of the abundance of shards to be processed which keeps the streaming multiprocessors busy. Second, unbalanced window sizes can cause intra-block divergence; however, due to similar reason, its impact is insignificant too. Finally, sparse graphs lead to small window sizes which cause GPU underutilization. This is mainly because most of the threads within the warp are idle when entries for the computation window are being processed by other threads. It makes shards representation on GPUs sensitive to window sizes; in particular, smaller window sizes lead to inefficient write back of updated values in the windows.

Concatenated Windows representation addresses the above issue. To avoid GPU underutilization induced by large sparse graphs, CW collocates computation windows. For a given shard i, a Concatenated Window CW_i is defined as a list of all computation windows W_{ij} *SrcIndex* elements, ordered by j. Concatenated Windows retains the original representation of shards, but separate out the *SrcIndex* entries to order them differently. *SrcIndex* entires for shard i in CW representation can be created by concatenating *SrcIndex* in all W_{ij} in G-Shards representation. Hence, a directed graph is represented as a set of shards, each of them associated with a separate *SrcIndex* array. Each shard is an ordered list of incoming edges where each edge is now represented by a three tuple: *SrcValue*, *EdgeValue*, and *DestIndex*. The set of shards is *Partitioned* and *Ordered* as described in the previous section. Note that by separating out *SrcIndex* array from the rest of the shard, we break the association between *SrcIndex* and *SrcValue* entries which is required to write back the updated

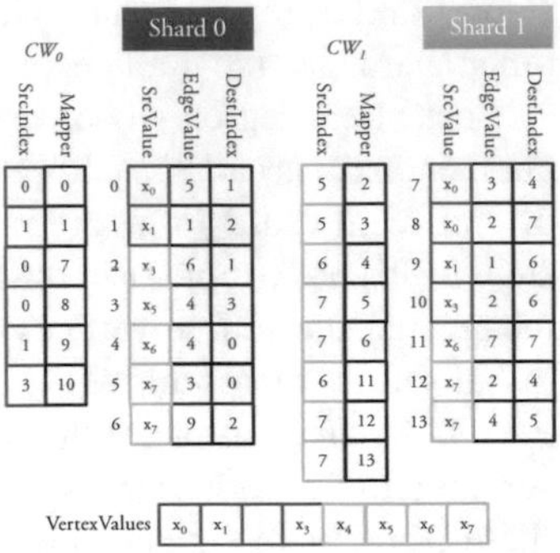

Fig. 5.4 *CW* representation of the graph in Fig. 5.1

SrcIndex	Mapper		SrcValue	EdgeValue	DestIndex		SrcIndex	Mapper		SrcValue	EdgeValue	DestIndex
				Shard 0							**Shard 1**	
			CW_0							CW_1		
0	0	0	x_0	5	1		5	2	7	x_0	3	4
1	1	1	x_1	1	2		5	3	8	x_0	2	7
0	7	2	x_3	6	1		6	4	9	x_1	1	6
0	8	3	x_5	4	3		7	5	10	x_5	2	6
1	9	4	x_6	4	0		7	6	11	x_6	7	7
3	10	5	x_7	3	0		6	11	12	x_7	2	4
		6	x_7	9	2		7	12	13	x_7	4	5
							7	13				

VertexValues: x_0 x_1 | | x_3 x_4 x_5 x_6 x_7

values. Therefore, to facilitate fast access of *SrcValue* entries using *SrcIndex* entries, an additional *Mapper* array is needed.

Processing graphs using the Concatenated Windows, similar to the G-Shards representation, takes four steps with a difference in 4th step (write-back stage). Using the entries in *SrcIndex* and *Mapper* arrays, the thread updates the corresponding *SrcValue* entries in the shard. By concatenating small windows to form larger set of entries, consecutive threads within the block continuously process consecutive entries, thus improving GPU utilization for large sparse graphs.

Figure 5.4 shows the Concatenated Windows representation for the graph shown in Fig. 5.2. As we can see, the *SrcIndex* columns are separately ordered compared to the rest of the shards. There are six entries in *SrcIndex* (shown in red) associated with *Shard-0* representing values from CW_0. The first two entries come from W_{00} and the rest come from W_{01}. The eight entries in *SrcIndex* (shown in green) are associated with *Shard-1* and represent values from CW_1. The first three come from W_{10} and the rest come from W_{11}.

5.3 SIMD-Efficient Task Decomposition Scheme

During graph processing, the neighbor visitation and reduction routine dominate the kernel's execution time; thus, its SIMD parallelization strategy for the GPU environment is of great importance. The main challenge of parallelization is *load balancing*. Vertices assigned to threads inside a warp can have different number of neighbors. In addition, they may or may not be active, which makes the set of to be visited adjacency lists in *C* array (*SrcIndex* array in Fig. 5.2) disjoint. While employing simpler static assignment approaches [2, 6] makes the kernel susceptible to load imbalance in presence of irregular graphs, we present an effective dynamic thread assignment technique [7] in this section that overcomes the above challenge. In this method, threads inside the warp iterate over a packed view of the neighbors belonging to different vertices. Therefore, in each round, each thread inside the SIMD group is

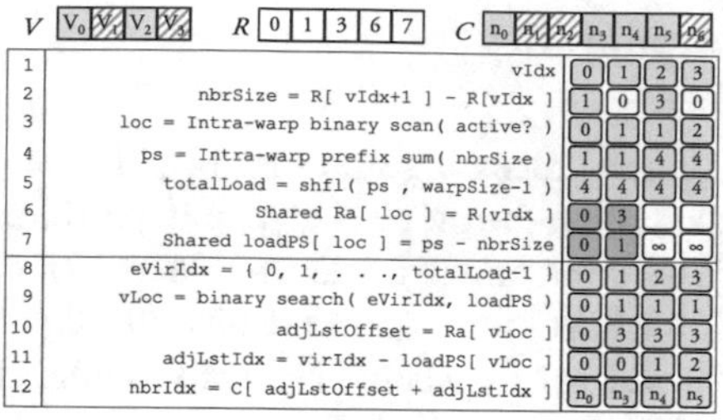

V | V₀ V₁ V₂ V₃ | R | 0 | 1 | 3 | 6 | 7 | C | n₀ n₁ n₂ n₃ n₄ n₅ |

#	Code	Values
1	vIdx	[0][1][2][3]
2	nbrSize = R[vIdx+1] - R[vIdx]	[1][0][3][0]
3	loc = Intra-warp binary scan(active?)	[0][1][1][2]
4	ps = Intra-warp prefix sum(nbrSize)	[1][1][4][4]
5	totalLoad = shfl(ps , warpSize-1)	[4][4][4][4]
6	Shared Ra[loc] = R[vIdx]	[0][3]
7	Shared loadPS[loc] = ps - nbrSize	[0][1][∞][∞]
8	eVirIdx = { 0, 1, . . ., totalLoad-1 }	[0][1][2][3]
9	vLoc = binary search(eVirIdx, loadPS)	[0][1][1][1]
10	adjLstOffset = Ra[vLoc]	[0][3][3][3]
11	adjLstIdx = virIdx - loadPS[vLoc]	[0][0][1][2]
12	nbrIdx = C[adjLstOffset + adjLstIdx]	[n₀][n₃][n₄][n₅]

Fig. 5.5 A simplified example demonstrating our effective dynamic thread assignment strategy. In the example, vertices V_0 and V_2 are active. Lines 1–7 are computations on vertices and lines 8–12 are computations on neighbors. V, C, and R correspond to *VertexValkues*, *SrcIndex*, and *InEdgeIdxs* arrays in Fig. 5.2 respectively

assigned to visit and process one neighbor, avoiding intra-warp load imbalance and leading to a sustained high warp execution efficiency.

Figure 5.5 illustrates the implementation details of dynamic thread assignment technique via an example. First, the number of neighbors that need to be visited is calculated for active vertices—threads inside the warp compute the intra-warp binary prefix sum using the vertex activeness binary predicate (line 3) so that threads with active vertices realize their relative location between similar threads inside the warp. Then threads perform an intra-warp inclusive prefix sum using the number of neighbors for active vertices (line 4), and shuffle the last lane's element to determine the total number of neighbors to visit (line 5). Two on-chip shared memory buffers are utilized to collect the row offset array elements and the exclusive prefix sum of number of neighbors both for active elements (lines 6 and 7).

Now, by knowing the total number of neighbors, threads can iterate over the loads without underutilization. The iteration index can be considered as a virtual neighbor index that has to be mapped to an actual neighbor inside the loop. Threads identify the active vertex location corresponding to their assigned virtual index using a binary search over the shared memory array containing the prefix sum of the number of neighbors (line 9), and figure out the adjacency list element index and offset inside C array (lines 10 and 11) using the information collected inside the shared memory. Therefore, threads find their assigned neighbors and continue the computation which involves a compute function with the neighbor content and reducing it with other neighbors. This technique essentially enables a one-to-one mapping with a thread and the neighbors of an active thread and therefore avoids thread underutilization. This issue can potentially introduce device underutilization, since, assuming similar GPU characteristics, all the GPUs have to wait for the GPU with the most active graph components to finish its job in an iteration. Essentially, GPU's load is mainly determined by the *activeness* of vertices.

5.4 Multi-GPU Scalability

In existing multi-GPU graph processing solutions [4], in order to distribute the computation across multiple GPUs, the graph is statically partitioned and each partition is assigned to one GPU. To maximize the allowable size of the input graph, a partition that is assigned to a GPU stays on the device throughout the iterative graph processing duration. However, this scheme can lead to inter-device load imbalance and GPU idling. This is because as the iterations go forward, the set of active vertices constantly change. In an iteration, a GPU that holds a partition of the graph and is assigned to process vertices in that particular partition may contain only a few active vertices and edges, while another GPU may end up with a considerably high number of active graph components.

Permissive partitioning [7] is a technique to mitigate this issue by allowing graph partitions that are being held by GPUs *overlap* as much as device memories allow—see example in Fig. 5.6. This enables us to dynamically slide the borders using which the GPUs determine the corresponding graph partitions they compute. The direction and the amount by which borders slide are determined on-the-fly during the iteration by observing the distribution of the active vertices in the bitmask. If a device in the current assignment has considerably higher number of active vertices and edges compared to other devices, the border that specifies its computation region is shrunk so that all devices have possibly equivalent or at least closer number of active graph components. Therefore, it provides a more balanced load and thus diminishes device underutilization.

Permissive partitioning has two phases: offline phase and online. In the offline phase, after performing an initial estimation of number of edges per device (as in static partitioning), the devices' available DRAM are queried and then the initial static vertex-range estimate is expanded to use the additional space available in the device memory. This essentially maximizes the usage of GPU's DRAM by storing as much of the graph data as possible.

For the online phase, which happens on-the-fly, at the beginning of each iteration of the iterative graph processing procedure, the number of active vertices in every fixed-sized group of vertices are counted by the GPU. Counting kernel utilizes the already existing bitmask and native population count primitive. The kernel writes the results directly into a preallocated host pinned buffer. After enqueuing all the commands to GPUs, the host waits for the events associated with the bitmask counting kernel. When this kernel ends, the host calculates an array containing the prefix sum of the element-wise multiplication outcome of active number of vertices and the

Fig. 5.6 An example of *permissive partitioning*. Partitions overlap in V_3, V_4, and V_5 and their edges

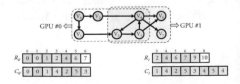

average number of edges per vertex in that section. When this array is created, the total approximate number of active edges can be found at the end of the array, and therefore, the fair amount of edges per device that balances load can be easily calculated. The host discovers the regions standing for the fair loads by binary searching the constructed array—which has a logarithmic time complexity. For each device, the host slides the border based upon the fair load value. If the load falls outside of the vertex region held by the device, the end of the region becomes the assignment border for the next iteration. Using permissive partitioning, at every iteration each device receives an approximately equal—or at least fairer—number of active edges.

5.5   Summary and Conclusion

In this article, we introduced challenges regarding high performance and scalable vertex-centric graph processing on GPUs and summarized some of the techniques that allow efficient graph computation using them. We first described graph representation formats including CSR, G-Shards, and CW. We discussed that G-Shards and CW enhance memory access efficiency compared to CSR at the expense of higher memory footprint. Next, we presented a dynamic task decomposition scheme for the threads within the warp that provides warp execution efficiency. Finally, we introduced a technique to scale the vertex-centric graph computation across multiple GPUs.

References

1. F. Khorasani, K. Vora, R. Gupta, L.N. Bhuyan, 2014. CuSha: vertex-centric graph processing on GPUs, in *Proceedings of the 23rd International Symposium on High-Performance Parallel and Distributed Computing (HPDC '14)* (ACM, New York, NY, USA), pp. 239–252. https://doi.org/10.1145/2600212.2600227
2. S. Hong, S.K Kim, T. Oguntebi, K. Olukotun, Accelerating CUDA graph algorithms at maximum warp, in *Proceedings of the 16th ACM Symposium on Principles and Practice of Parallel Programming (PPoPP '11)* (ACM, New York, NY, USA, 2011), pp. 267–276. https://doi.org/10.1145/1941553.1941590
3. A. Gharaibeh, L.B. Costa, E. Santos-Neto, M. Ripeanu, A yoke of oxen and a thousand chickens for heavy lifting graph processing, in *2012 21st International Conference on Parallel Architectures and Compilation Techniques (PACT)* (Minneapolis, MN, 2012), pp. 345–354
4. F. Khorasani, R. Gupta, L.N. Bhuyan, Scalable SIMD-efficient graph processing on GPUs, in *2015 International Conference on Parallel Architecture and Compilation (PACT)* (San Francisco, CA, 2015), pp. 39–50. https://doi.org/10.1109/PACT.2015.15
5. A. Kyrola, G.E. Blelloch, C. Guestrin, textitGraphchi: Large-scale graph computation on just a pc (USENIX, 2012)
6. P. Harish, P.J. Narayanan, Accelerating large graph algorithms on the GPU using CUDA. HiPC **7**, 197–208 (2007)
7. F. Khorasani, *High Performance Vertex-Centric Graph Analytics on GPUs* (University of California, Riverside, PhD diss., 2016)

Chapter 6
Security Challenges of Networked Control Systems

Arman Sargolzaei, Alireza Abbaspour, Mohammad Abdullah Al Faruque, Anas Salah Eddin and Kang Yen

6.1 Introduction

The expanded demand on critical infrastructure facilities has resulted in an increased dependency on automated industrial systems. This dependency is clearly visible in several sectors such as energy, transportation, and health care. Consequently, complex control systems are being utilized to improve the performance of these automation systems while reducing the number of human operators. Specifically, the use of network control systems (NCSs) to monitor and control a wide variety of subsystems, sensors, and actuators has been greatly increased. This increased adoption and the increase in design complexity has led to security challenges, which resulted in an interest in studying the resilience and reliability of such systems under different attack scenarios.

A. Sargolzaei (✉)
Department of Electrical and Computer Engineering, Florida Polytechnic University, Lakeland, FL, USA
e-mail: a.sargolzaei@gmail.com

A. Abbaspour · K. Yen
Department of Electrical and Computer Engineering, Florida International University, Miami, FL, USA
e-mail: alireza.abaspour@gmail.com

K. Yen
e-mail: kang.yen@fiu.edu

M. A. Al Faruque
Department of Electrical and Computer Science, University of California, Irvine, CA, USA
e-mail: alfaruqu@uci.edu

A. Salah Eddin
Department of Electrical and Computer Engineering, California State Polytechnic University, Pomona, CA, USA
e-mail: asalaheddin@cpp.edu

© Springer International Publishing AG, part of Springer Nature 2018
M. H. Amini et al. (eds.), *Sustainable Interdependent Networks*,
Studies in Systems, Decision and Control 145,
https://doi.org/10.1007/978-3-319-74412-4_6

Automation of Industrial Control Systems (ICSs) is achieved by employing a wide variety of subsystems including load frequency control (LFC), supervisory control and data acquisition (SCADA), distributed control systems (DCSs), process control systems (PCSs), and programmable logic controllers (PLCs). The diversity of these subsystems makes them especially vulnerable to several attack vectors. A quick survey of recent news and literature reports shows the severe and adverse effects of such attacks from financial and security perspectives. The following is a non-comprehensive list of some notorious attacks: the attack on Marrochy water bridge [1], the attack on a nuclear plant using the StuxNet virus [2], this attacks targeted a SCADA system in the plant, the capturing of a reconnaissance unmanned aircraft, RQ-170 [3], the attack on anti-lock braking system (ABS) [4]. This shows that despite the increased attention to improving the security of control systems there is still a lot of work to be done, explicitly in the development of detection, prevention, and recovery algorithms [5, 6]. In the following subsections, we expand on some of the recent attacks. We focus on attacks targeting sensors and actuators since they are the backbone of NCSs; therefore, we pay special attention to power systems, unmanned vehicles, and other miscellaneous NCSs.

6.1.1 Attacks on the Power Systems

Failure of the power grid may lead to catastrophic consequences [7–10], incidents like the wide power blackout of 2003 in northeastern United States and Canada [11] and the blackout of 2012 in India [12] have highlighted the importance of a reliable and resilient power system. In 2015, the effects of the first known cyber-attack on a power distribution gird was observed in Ukraine, more than 230 thousand people living in the affected areas lost power [13]. Here, we focus on the security of NCSs used in power systems and review possible attack prevention solutions.

In [14], Farraj et al. introduced a resilient controller that could tolerate disturbances in the power grid. The controller was based on linearizing feedback of the synchronous generators' transient stability. The proposed controller relied on receiving timely information from the phase measurement unit (PMU) in the power grid parts and it had the advantage of being tunable in case a disturbance occurred.

Some attacks interrupt the controllability and observability of a control system. In [15], Li et al. investigated the controllability and observability of an NCS under such attacks. Specifically, they investigated attacks on the actuator system of a 9-bus power grid system. The *controllability* under adversarial attacks was defined as the possibility of sending an accurate control signal to the attacked system and stabilizing it. Similarly, observability was defined as the ability to correctly reconstruct the system states. The controller and observer are designed based on linear model of the system.

Nonlinear estimation methods are more complicated in their structure than linear methods; however, they are more desirable due to their accuracy under severe disturbances and uncertainties. In [16], Hu et al. introduced a secure state estimation method for nonlinear power systems. They used an l_1 optimization algorithm in a recursive mode to reduce the computational complexity and the time of state estimation. Despite the fact that this method is accurate, it is not able to detect attacks such as time delay switch (TDS) attack.

In [17–19], Sargolzaei et al. introduced a TDS attack and demonstrated the vulnerability of an NCS to this kind of attack scenario. They performed simulations on a power distribution system and showed the destructive effect of this attack. To recover from the TDS attack, they used a linear time delay estimator and a model reference control approach. The model was enhanced by an indirect supervisor and an enhanced least-mean-square minimization algorithm. In another effort, they designed a novel detection mechanism to estimate TDS attack and update the transmitter to address the negative impacts of this attack [19, 20].

6.1.2   Attacks on Unmanned Systems

Unmanned systems (i.e., unmanned aerial vehicles and self-driving cars) are controlled by an autopilot system that depend on a number of sensors and actuators linked through a network. Similar to the other NCSs discussed above, the growth of these systems should be associated with increased security measures. Several studies have focused on attack detection algorithms. For example, in [21], Abbaspour et al. introduced an artificial neural network (ANN) based observer to detect faults in the actuators and the sensors of an unmanned aircraft. A nonlinear observer and an ANN were used where the learning weights were updated by an extended Kalman filter (EKF). The filter was designed to detect anomalies in the actuators and sensors of the aircraft. The work was further extended in [22] to detect FDI attacks in the aircraft sensors. Similarly, in [23], Rana introduced a Kalman filter-based algorithm to detect cyber-attacks on the NCS of smart vehicle sensors. The estimator was designed to evaluate the sideslip angle of the vehicle in the presence of a cyber-attack on the sensors. Additionally, several cyber-attacks on the communication links of the adaptive cruise control of a self-driving car were investigated in [24]; Noei et al. introduced a fuzzy algorithm to detect the attack and a PID controller to avoid collisions. Ghanavati et al. [25] used partial differential equations (PDE) models to study the effect of a cooperative cyber-attack on networked autonomous vehicles on highways.

6.1.3   Miscellaneous Applications

Security of NCSs is not limited to the applications discussed above; for example, attackers have targeted the natural gas industry. The Russian Interior Ministry observed attacks on their gas flow regulator systems in 2000 [26]. Similarly, NCSs in transportation systems have been targeted. In 2008, a teenager modified a TV remote to control tramway tracks, the ramifications were devastating resulting in four derailments and several injuries [27]. Moreover, many biomedical devices have control systems. Security of such devices is critical as attacks might lead to fatalities. The US Department of Homeland Security (DHS) has recently reported some threats affecting around 300 medical devices. Particularly vulnerable targets are networked and wireless devices such as pacemakers [28–30] and infusion pumps [31]. Additionally, wastewater management facilities utilize control systems to treat wastewater and remove harmful particles before it can be recycled to clean water. These control systems have also been the subject of reported attacks. In January 2000, a sewage control system in Australia was targeted and attacked, its consequences remained for several months and led to flooding of the grounds in a nearby park, a hotel and a river with a million liters of untreated sewage [1].

Several studies have emerged in response to the cyber-attacks discussed above. Satchidanandan and Kumar [32] used a watermarking approach to find malicious data in sensor measurements. They relied on injecting a known excitation signal with a known expected response from the actuator to the system. The accuracy of sensor measurements was then determined by comparing the observed response with the expected one. This approach was based on a very accurate model of the dynamic system and could not be extended to models with uncertainties. In [33], Pajic et al. used an l_0 state estimator to recover from cyber-attacks in the presence of bounded noise. The system state was successfully estimated during an attack in the presence of significant system noise. It was also concluded that when the number of attacks is less than the observer states, an attacker cannot use uncertainties and noise models to inject unbounded estimation error. Ding et al. [34] used a χ^2 detection scheme which uses a Kalman filter estimation technique, the maximum number of attacks and the probability of an attack was determined by recursively solving the Riccati-like difference equations. Detection of cyber-attacks in an event triggering consensus control algorithm was investigated in [35], when a violation is detected in its triggering condition, the controller updated communications among its agents. This chapter is structured as follows: Sect. 6.2 provides a general definition of NCSs and security challenges related to them. Section 6.3 includes a general mathematical formulation of NCSs under attack. Section 6.4 proposes a simple technique to overcome the negative impacts of TDS and FDI attacks. Thereafter, a case study using a power system as an example is presented. We show the effects of cyber-attacks on different parts of the NCS. We also discuss testing and verification of the accuracy of our technique. The chapter is finalized with a brief summary and conclusions.

6.2 Networked Control Systems (NCS)

NCSs can be classified into several different types based on their design and topology. A typical NCS is a basic control system while the feedback process is closed through a communication channels [36, 37]. In this topology, the information and data of the control system can be sent to other nodes/systems outside the system. Figure 6.1 is a

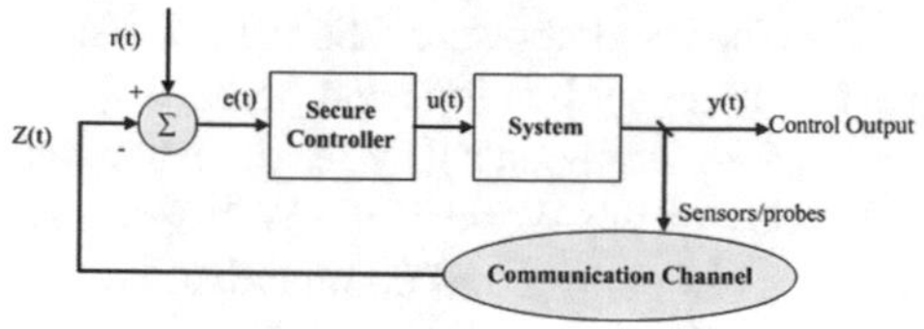

Fig. 6.1 Typical networked control systems

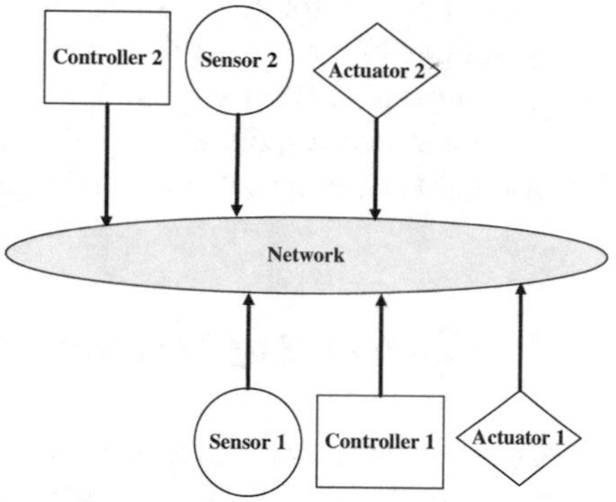

Fig. 6.2 Shared network structural networked control systems

diagram of such an NCS in which sensors measure the output signals and transmit the sensor measurements to the controller through communication channels. Afterward, the controller checks the measured parameters against reference values and generates the appropriate control signal.

Figure 6.2 illustrates a second type of NCSs, where the system depends on a shared network design. In such systems, controllers, sensors, and actuators share the same network. Although this type of NCSs is cost efficient, it is more vulnerable to cyber-attacks. Furthermore, this type might not be pertinent to time-sensitive applications.

In certain applications, NCSs can be connected to each other and/or controlled by a centralized networked control system (CNCS) as shown in Fig. 6.3. They can also receive the reference signal through a network from a supervisory center.

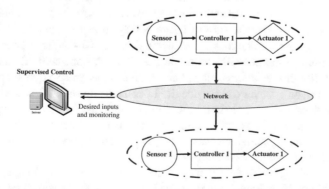

Fig. 6.3 Centralized/hierarchical structural networked control systems

As discussed above, NCSs are pervasively used in industry and play an integral part in many applications and fields [36]. NCSs can be called industrial control networks, industrial control systems, distributed control systems, command and control networks, or the World Wide Web systems, based on their applications. Technical advances in control and communication have made it possible to control and monitor several systems over a wide area network. This allows engineers to monitor, configure, and control systems from a distance easier, it also allows to coordinate the operations of a complex distributed infrastructure with high efficiency. Due to the communication connectivity of NCSs and the fact that they are used to control critical infrastructures, the security of NCSs is very relevant and the next subsection is a discussion of some security challenges related to NCSs.

6.2.1 Challenges of Networked Control Systems

The catastrophic consequences of a potential cyber-attack on NCSs [38, 39] make the security and reliability of NCSs as a main priority for the control system designers. Despite the significant improvement in the security level of the NCSs in recent years, many acknowledged cyber-attacks have been reported which shows the vulnerability of these control system against multiple attacks and disturbances. By reviewing the related literature, it can be concluded that a major attention is needed for the improvement of security of the NCSs. The key problem with the conventional control system is that they are designed based on the assumption of normal data transfer without any intrusion. In these kinds of controllers, any attack on any part of a control system such as sensors, actuators, or communication links may degrade the system performance and make the system unstable.

The modernization of the NCSs has been continued in the past decades by the introduction of new communication and processing technologies for online control and monitoring of the system. However, these modernizations lead to vulnerability of these systems to cyber-attacks in which a potential incident can have a huge adverse impact on people's lives and economy. For instance, consider the US power networks which operate with a supervisory control and data acquisition (SCADA) system. SCADA systems are applied to control large network systems which include multiple subsystems and sites. Despite various protection systems designed for SCADA systems, several cyber-attacks have been reported [11, 12, 40–42]. Moreover, replacing private communication networks with open communication standards will increase the vulnerability of the process control and SCADA systems to cybersecurity risks [43]. In general, an attacker can intrude the IT infrastructure of ICSs, and get access to various sensor and control signals, and inject malicious data to disrupt and sabotage the system. As an example, consider an intruder that by sending malicious data in the feedback system increases a load on particular power transformer, or shut down a section of a control system [44, 45].

Another challenge with the security of NCSs is the cost of redesign and collaboration problem. If resilience is not considered at the early stages of design, the cost of redesign after or under a cyber-attack increases significantly. Therefore, most industries prefer to address this issue at the application layer and modify the network and communication side rather than focusing on the resilient design of both control and communication components. Moreover, it was mentioned earlier that NCSs are used in different critical infrastructure facilities, collaboration and communication between the different sectors is not always attainable. As a result, responsive modification of the design is not feasible. Furthermore, due to the sensitivity of the data and applications, these facilities only work with high-level security clearance contractors and companies to repair and redesign their systems. Consequently, most of the proposed protection techniques cannot be tested in real-world applications.

Another challenge is the randomness of attacks. Researchers do not know when and where the attack can happen nor can they predict the type of attack, researchers need a new vision and a strategy of defense against cyber-attacks [46, 47].

Although there are many other challenges in this domain, we only summarized some of the important ones. The final challenge is constraints of resources. Identifying the vulnerabilities of safety-critical infrastructure, the related risks, and available countermeasures have attracted significant academia and industry attention. Despite the limited resources available, the current efforts are dedicated to the development of secure hardware and software systems and protocols to predict, prevent, and respond to the common types of cyber threats. Speed, cost, and energy efficiency of hardware used in NCSs are some of the limitations [48].

Design and implementation of a robust, secure, and reliable NCS requires an understanding of the vulnerable assets, common threats, prediction and estimation strategies, and proper actions in response to cyber-attacks, failures, and faults. The goal of the next section is to provide a mathematical formulation for an NCS under attacks.

6.3   Mathematical Formulation of NCSs Under Attack

The following mathematical formulation focuses on a simple NCS. As an example, Fig. 6.4 shows a single plant consisting of different ith areas, and it also shows that an adversary could attack the system at different points: the control and supervisory center (point 1), communication links (point 2 and 4), controller and actuator (point 3), and wireless sensors (point 4). The following mathematical formulation describes the system with an output feedback:

$$\begin{cases} \dot{X}(t) = F(t, U(t), X(t)) \\ Y(t) = G(X(t)) \end{cases} \tag{6.1}$$

$F = \{f_1, f_2, \dots, f_i\}$ and $G = \{g_1, g_2, \dots, g_i\}$ are functions that describe the plant behavior and plant output. $X = [x_1, x_2, \dots, x_i]^T$, $U = [u_1, u_2, \dots, u_i]^T$, $Y = [y_1, y_2, \dots, y_i]^T$ are the state, input, and output vectors of the multi-area NCS. Each of the ith

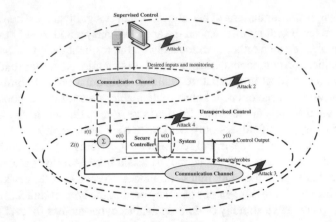

Fig. 6.4 A networked control system under attack

areas can have several state variables. For example, an ith NCS can have N state variable and can be denoted by $x_i(t) = [x_{i,1}, x_{i,2}, \ldots, x_{i,N}]^T$.

The measured output signal is transmitted to the controller through a communication link and can be described by:

$$Z(t) = H(Y(t), t) \tag{6.2}$$

where $H = \{h_1, h_2, \ldots, h_i\}$ is the communication methodologies utilized by the different links.

Similarly, $Z(t) = [z_1, z_2, \ldots, z_i]^T$ describes the measured vectors received by the system, each of which has an appropriate dimension. The controller can be presented as

$$U(t) = M(t, Z(t), r(t)) \tag{6.3}$$

where $U(t)$ is the input vector, $r(t) = H(r_{or}, t)$ is the reference signal, r_{or} is the original transmitted reference signal before sending through the communication channel, and $M = \{m_1, m_2, \ldots, m_i\}$ is a function that describes the controller.

NCSs can be attacked by altering any of their parts. A general attack can be described as a function that affects different parts of NCSs:

$$(\hat{x}_i, \hat{y}_i, \hat{f}_i, \hat{g}_i, \hat{z}_i, \hat{r}_i, \hat{U}_i, \hat{t}) = \pi(x_i, y_i, f_i, g_i, z_i, u_i, r_i, t) \tag{6.4}$$

π is an attack function that can model the injection of TDS, DoS, FDI, and other types of attacks on the different NCS' components.

The TDS attack function can be defined as:

$$\begin{cases} \pi(\gamma(t)) = \hat{\gamma}(t) = \gamma(t - \tau(t)) \\ \text{or} \\ \pi(t) = \hat{t} = t - \tau(t) \end{cases} \tag{6.5}$$

where τ is a positive time-varying random value.

Similarly, the FDI attack function can be defined as:

$$\pi(\gamma(t)) = \hat{\gamma}(t) = \gamma(t) + \alpha(t) \tag{6.6}$$

where α is a random variable, which can be positive or negative.

Finally, the π function for DoS is defined as follows:

$$\begin{cases} \pi(\gamma(t)) = \hat{\gamma}(t) = \alpha \\ \text{or} \\ \pi(t) = \hat{t} = \infty \end{cases} \tag{6.7}$$

6.4   Comprehensive Attack Detection and Recovery

Following the mathematical formulation above, we propose a solution focused on attacks targeting the feedback line. We utilize an attack detector based on an adaptive communication technique and a recovery system that compensates the negative attack effects on the feedback line; specifically, the negative effects of TDS and FDI attacks. Since collaboration and the cost of controller redesign are one of the prohibitive challenges in NCSs security, we propose an effective and simple method that does not require redesign. The attack detector is based on previous work [17, 21, 49], where it can detect TDS and FDI attacks simultaneously. Figure 6.5 is a diagram of the proposed solution, and it shows that the system includes: (1) A data transmitter (Tx), which adaptively allocates transmission channels and sends the sensor measurement signals. (2) An observer that estimates the current plant states. (3)

Fig. 6.5 Block diagram of proposed method

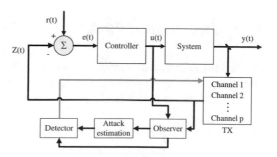

A TDS and FDI estimator to continuously estimate the time delays and faults on the channels. (4) A TDS and FDI detector to determine if the delays are more than an acceptable threshold, then issue commands to the transmitter to request a new channel. (5) A controller to control the system.

The functionality of the attack detector can be described by the following equation:

$$D(t) = \begin{cases} 1 & (\hat{\tau} > \tau_{stable}) \quad \text{or} \quad (\hat{\alpha} > \alpha_{stable}) \\ 0 & \text{otherwise} \end{cases} \tag{6.8}$$

where $\hat{\tau}$ and $\hat{\alpha}$ are the estimated time delay and the rate of false data injected in the system, respectively. τ_{stable} and α_{stable} are the maximum allowable time delay and the fault that the system can tolerate before being unstable or inefficient. More information about the channel allocation technique can be found in the literature [50].

6.5 A Case Study of an NCS Under Attack

To illustrate the effect of common attacks on different parts of an NCS, a centralized load frequency controller (LFC) with two power areas is selected. The power areas are connected through communication channels to the centralized LFC, which sends control signals to the plant and receives state signals through sensor measurements. One of the crucial aspects of power networks is using efficient algorithm to manage the load. Amini et al. [10] introduced an efficient multi-agent-based framework to integrate distributed generations and enable demand response programs. Their method comprehensively covers the details of smart power systems to reduce the load demand and help both costumers and utilities to optimize their costs [10]. To take advantage from this algorithm and others, the NCS should be secure. The attacks can target the LFC control signals, feedback measurement path, or reference signals by jamming the communication channels. It should be noted that an adversary can also attack a supervisory center, each power area, actuators, and other components of the centralized NCS.

6.5.1 Mathematical Model

The following is a mathematical model and brief description summarizing an interconnected multi-area power system. Detailed explanation can be found in [17, 51]. Consider the linear approximation model of the system:

$$\begin{cases} \dot{X}(t) = AX(t) + BU(t) + D\Delta P_l, \quad X(0) = X_0 \\ Y(t) = CX(t) \end{cases} \tag{6.9}$$

where $U(t)$ and $X(t) = [x_1, x_2, \ldots, x_N]^T$ are the input and state vectors, respectively. The ΔP_l is the load deviation. The system states for the ith power area are $x_i = [\Delta f^i, \Delta P_g^i, \Delta P_{tu}^i, \Delta P_{pf}^i, e^i]^T$ where $\Delta f^i, \Delta P_g^i, \Delta P_{tu}^i, \Delta P_{pf}^i, e^i$ are frequency deviation, generator power deviation, turbine valve position, power flow of the tie-line and control error, respectively. The control error of the power area is $e^i(t) = \int_0^t \beta^i \Delta f^i dt$ where β^i is the frequency bias factor. The initial state condition is denoted by X_0 and C is a unit matrix.

A and B are constant matrices with appropriate dimensions, both can be determined by

$$A = \begin{bmatrix} A_{1,1} & A_{1,2} & \cdots & A_{1,N} \\ A_{2,1} & A_{2,2} & \cdots & A_{2,N} \\ \vdots & \vdots & \vdots & \vdots \\ A_{N,1} & A_{N,2} & \cdots & A_{N,N} \end{bmatrix} \tag{6.10}$$

$$B = diag\{ \begin{bmatrix} B_1 & B_2 & \cdots & B_N \end{bmatrix}^T \} \tag{6.11}$$

$$D = diag\{ \begin{bmatrix} D_1 & D_2 & \cdots & D_N \end{bmatrix}^T \} \tag{6.12}$$

N is the number of power areas.

In an ideal case, $Z(t) = X(t)$; the control signal $U(t)$ can be expressed by

$$U(t) = -KZ(t) \tag{6.13}$$

where K is a constant matrix that can be calculated using an optimal control method or through pole placement method.

The matrices $A_{i,j}$, B_i, and D_i for $i = 1, 2, \ldots, N$ can be calculated as follows:

$$A_{i,i} = \begin{bmatrix} -\mu_i/J_i & 1/J_i & 0 & -1/J_i & 0 \\ 0 & -1/T_{tu,i} & 1/T_{tu,i} & 0 & 0 \\ -1/\omega_i T_{g,i} & 0 & -1/T_{g,i} & 0 & 0 \\ \sum_{i=j,j=1}^{2} 2\pi T_{i,j} & 0 & 0 & 0 & 0 \\ \beta_i & 0 & 0 & 0 & 1 \end{bmatrix} \tag{6.14}$$

$$A_{i,j} = \begin{bmatrix} 0 & 0 & 0 & 0 & 0 \\ 0 & 0 & 0 & 0 & 0 \\ 0 & 0 & 0 & 0 & 0 \\ -2\pi T_{i,j} & 0 & 0 & 0 & 0 \\ 0 & 0 & 0 & 0 & 0 \end{bmatrix} \tag{6.15}$$

$$B_i = \begin{bmatrix} 0 & 0 & 1/T_{g,i} & 0 & 0 \end{bmatrix}^T \tag{6.16}$$

$$D_i = \begin{bmatrix} -1/J_i & 0 & 0 & 0 & 0 \end{bmatrix}^T \tag{6.17}$$

where ω_i, J_i, $T_{g,i}$, μ_i, and $T_{tu,i}$ are the speed-droop coefficient, the generator's moment of inertia, the governor time constant, damping coefficient, and the ith power area's turbine time constant. $T_{i,j}$ is the stiffness constant between the ith and jth power areas.

To illustrate the effect of attacks on NCSs and explain the countermeasures in details, we selected two power areas $N = 2$. Substituting the mentioned parameters with others obtained from a previous study [17] we obtain:

$$A_{1,1} = \begin{bmatrix} -0.15 & 0.1 & 0 & -0.1 & 0 \\ 0 & -5 & 5 & 0 & 0 \\ -166.66 & 0 & -8.33 & 0 & 0 \\ 1.24 & 0 & 0 & 0 & 0 \\ 21.5 & 0 & 0 & 0 & 1 \end{bmatrix} \tag{6.18}$$

$$A_{2,2} = \begin{bmatrix} -0.083 & 0.083 & 0 & -0.083 & 0 \\ 0 & -2.22 & 2.22 & 0 & 0 \\ -11.11 & 0 & -5.55 & 0 & 0 \\ 1.24 & 0 & 0 & 0 & 0 \\ 21.5 & 0 & 0 & 0 & 1 \end{bmatrix} \tag{6.19}$$

$$A_{1,2} = A_{2,1} = \begin{bmatrix} 0 & 0 & 0 & 0 & 0 \\ 0 & 0 & 0 & 0 & 0 \\ 0 & 0 & 0 & 0 & 0 \\ -1.24 & 0 & 0 & 0 & 0 \\ 0 & 0 & 0 & 0 & 0 \end{bmatrix} \tag{6.20}$$

$$B_1 = \begin{bmatrix} 0 & 0 & 8.33 & 0 & 0 \end{bmatrix}^T \tag{6.21}$$

$$B_2 = \begin{bmatrix} 0 & 0 & 5.55 & 0 & 0 \end{bmatrix}^T \tag{6.22}$$

The MATLAB function 'lqrd' was used to select the controller gain $K = [k_1 k_2]^T$, where 'lqrd' is based on a linear quadratic (LQ) regulator for a continuous plant.

$$K = \begin{bmatrix} 0 & 0 & 0 & 0 & 0 & 0 & 0 & 0 & 0 & 0 \\ 0 & 0 & 0 & 0 & 0 & 0 & 0 & 0 & 0 & 0 \\ 49.12 & 1.01 & 0.77 & -4.22 & 5.05 & -0.14 & -0.0015 & -0.0001 & 0.48 & -0.29 \\ 0 & 0 & 0 & 0 & 0 & 0 & 0 & 0 & 0 & 0 \\ 0 & 0 & 0 & 0 & 0 & 0 & 0 & 0 & 0 & 0 \\ 0 & 0 & 0 & 0 & 0 & 0 & 0 & 0 & 0 & 0 \\ 0 & 0 & 0 & 0 & 0 & 0 & 0 & 0 & 0 & 0 \\ 0.04 & 0.0001 & -0.0001 & -0.17 & -0.32 & 76.19 & 2.07 & 0.896 & -6.08 & 6.49 \\ 0 & 0 & 0 & 0 & 0 & 0 & 0 & 0 & 0 & 0 \\ 0 & 0 & 0 & 0 & 0 & 0 & 0 & 0 & 0 & 0 \end{bmatrix} \tag{6.23}$$

The R and Q are selected as $100 * diag(ones(1, 5 \times N))$.

Fig. 6.6 All states of the two-area power system under normal operation

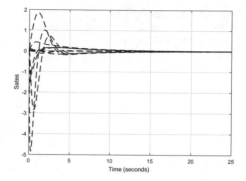

6.5.2 Simulation Results for a System Under Attack

A TDS attack targets the communication links between sensors and actuators, and the attack disrupts the reference signal making the system inefficient and potentially unstable. In this case study, the TDS attack cannot affect the system's stability nor its efficiency, because an optimal controller design was used to follow a zero-referenced signal. Moreover, the system was kept between specified boundaries using a stabilizer. Figure 6.6 shows all ten states for a system under normal operation.

6.5.2.1 Attacks on Sensor Measurements

TDS attacks: Considering a TDS attack on the first power area, where the attack targets the sensory measurements (i.e., the third state of the system). The third state of each power area would be transmitted to a centralized controller through the communication channel. Figure 6.7 shows the third state of the system under: normal operation (solid line), TDS attack 1 (dashed line), TDS attack 2 (dotted line), and TDS attack 3 (dot-dashed line). The different TDS attacks were generated by chang-

Fig. 6.7 Third state (Valve position of the turbine system) of the system under TDS attacks

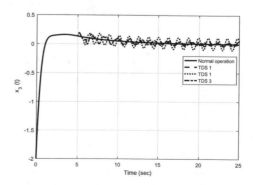

Fig. 6.8 Third state (Valve position of the turbine system) of the system under FDI attacks

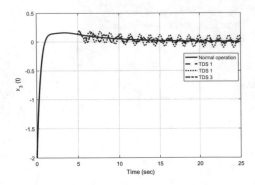

ing the start time and amount of injected delay. The specifications of each simulated TDS attack are:

$$\text{TDS 1:} \begin{cases} t_s = 10\,\text{s} \\ \tau = 0.7\,\text{s} \end{cases} \tag{6.24}$$

$$\text{TDS 2:} \begin{cases} t_s = 5\,\text{s} \\ \tau = 0.7\,\text{s} \end{cases} \tag{6.25}$$

$$\text{TDS 3:} \begin{cases} t_s = 5\,\text{s} \\ \tau = 1\,\text{s} \end{cases} \tag{6.26}$$

where t_s and τ are the attack start time and amount of injected delay, respectively.

FDI attacks: Similarly, the third state of the system can be targeted by an FDI attack. Figure 6.8 shows the system under three different FDI attacks as summarized bellow:

$$\text{FDI 1:} \begin{cases} 0.5 sin(5t) & t \geqslant 10 \\ 0 & \text{Otherwise} \end{cases} \tag{6.27}$$

$$\text{FDI 2:} \begin{cases} 0.5 & t \geqslant 10 \\ 0 & \text{Otherwise} \end{cases} \tag{6.28}$$

$$\text{FDI 3:} \begin{cases} 0.5 & t \geqslant 5 \\ 0 & \text{Otherwise} \end{cases} \tag{6.29}$$

Fig. 6.9 Third state of the system while first power area's control signal is under TDS attacks

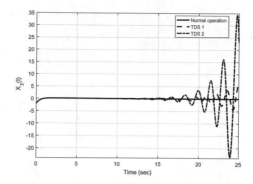

6.5.2.2   Attacks on the Control Signal/Actuator

TDS attacks: Control signals determined by the controller are sent to the first power area via the communication links. As in the attacks described above, these links are vulnerable to TDS attacks. Figure 6.9 shows the third state of a system under normal conditions (solid line), TDS attack 1 (dashed line), TDS attack 2 (dash-dotted line). TDS attacks 1 and 2 started at time $t_s = 10$ s with delay amounts of $\tau = 0.4$ and $\tau = 0.5$, respectively.

6.5.3   *Simulation Results for the Attack Detection and Recovery*

A Leuenberger observer was utilized using design recommendations in [52]. Similarly, the FDI and TDS attack detectors were designed based on previous work [49] and [17]. The FDI detector was designed using an artificial neural network with five hidden layers. As in the attack scenarios described above, the system was tested and simulated assuming the third state of the system was the target of attack. Additionally, the number of communication channels available for the transmission signals was restricted to five. It was also decided that an adversary has the ability to detect the change in channels and the ability to change the attack parameters one after another with the following specifications:

$$\text{Attack 1 on channel 1:} \begin{cases} 0.5sin(5t) & t \geqslant 0.1 \\ 0 & \text{Otherwise} \end{cases} \tag{6.30}$$

$$\text{Attack 2 on channel 2:} \begin{cases} t_s = 3\,\text{s} \\ \tau = 0.7\,\text{s} \end{cases} \tag{6.31}$$

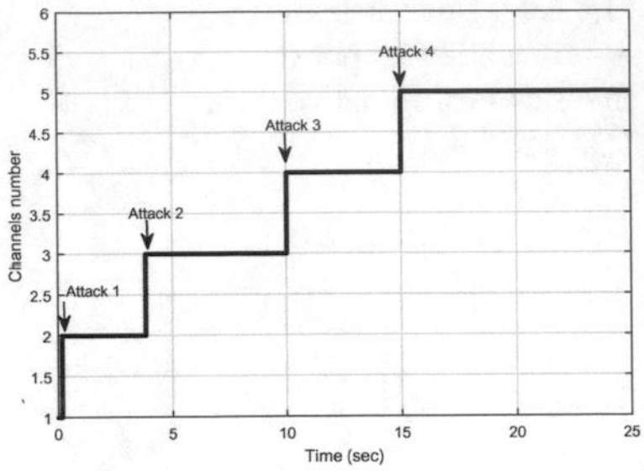

Fig. 6.10 Attack detection
on different channels and
requesting new channel

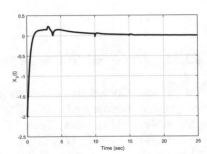

Fig. 6.11 Third state of the
system under attacks with
the proposed method

$$\text{Attack 3 on channel 3:} \begin{cases} 0.5 & t \geqslant 10 \\ 0 & \text{Otherwise} \end{cases} \qquad (6.32)$$

$$\text{Attack 4 on channel 4:} \begin{cases} 0.5 & t \geqslant 15 \\ 0 & \text{Otherwise} \end{cases} \qquad (6.33)$$

Figure 6.10 shows that the attack detector can accurately identify attacks and
effectively request a change in the communication channel. The effect of the control system including attack detection and recovery of third state is illustrated in
Fig. 6.11.

6.6 Summary and Conclusion

This chapter focused on security challenges facing NCSs such as collaboration
issues, the cost of redesign, hardware and software limitation, and many others. In
addition, a general mathematical model for NCSs under attacks was presented. A
practical case study was also discussed, where a centralized LFC with two power
areas was selected as an NCS under attack. TDS and FDI attacks were used to
show the negative impacts of such attacks. A detection and recovery system was also

presented levering state estimation and adaptive channel allocation techniques. Simulation results have demonstrated the capabilities of the detection and recovery system in the face simultaneous TDS and FDI attacks.

References

1. J. Slay, M. Miller, Lessons learned from the maroochy water breach. Crit. Infrastruct. Prot. 73–82 (2007)
2. R. Langner, Stuxnet: dissecting a cyberwarfare weapon. IEEE Secur. Priv. **9**(3), 49–51 (2011)
3. S. Peterson, P. Faramarzi, Iran hijacked US drone, says Iranian engineer. Christ. Sci. Monit. **15** (2011)
4. Y. Shoukry, P. Martin, P. Tabuada, M. Srivastava, Non-invasive spoofing attacks for anti-lock braking systems, in *International Workshop on Cryptographic Hardware and Embedded Systems* (Springer, 2013), pp. 55–72
5. S. Sargolzaei, M. Cabrerizo, A. Sargolzaei, S. Noei, M. Adjouadi, Epilepsy, a cyberattack on brains networked control system, in *2016 15th IEEE International Conference on Machine Learning and Applications (ICMLA)* (IEEE, 2016), pp. 622–625
6. A. Sargolzaei, M. Abdelghani, K.K. Yen, S. Sargolzaei, Sensorimotor control: computing the immediate future from the delayed present. BMC Bioinform. **17**(7), 245 (2016)
7. J. Hare, X. Shi, S. Gupta, A. Bazzi, Fault diagnostics in smart micro-grids: a survey. Renew. Sustain. Energy Rev. **60**, 1114–1124 (2016)
8. S. Amin, A.A. Cárdenas, S. Sastry, Safe and secure networked control systems under denial-of-service attacks, in *HSCC*, vol. 5469 (Springer, 2009), pp. 31–45
9. M.H. Amini, K.G. Boroojeni, T. Dragičević, A. Nejadpak, S. Iyengar, F. Blaabjerg, A comprehensive cloud-based real-time simulation framework for oblivious power routing in clusters of dc microgrids, in *2017 IEEE Second International Conference on DC Microgrids (ICDCM)* (IEEE, 2017), pp. 270–273
10. M.H. Amini, B. Nabi, M.-R. Haghifam, Load management using multi-agent systems in smart distribution network, in *2013 IEEE Power and Energy Society General Meeting (PES)* (IEEE, 2013), pp. 1–5
11. S. Gorman, Electricity grid in US penetrated by spies. Wall Str. J. **8** (2009)
12. H. Pidd, India blackouts leave 700 million without power. Guardian **31** (2012)
13. K. Zetter, Inside the cunning, unprecedented hack of Ukraines power grid. Wired (2016)
14. A. Farraj, E. Hammad, D. Kundur, A cyber-enabled stabilizing control scheme for resilient smart grid systems. IEEE Trans. Smart Grid **7**(4), 1856–1865 (2016)
15. Y. Li, J. Wu, S. Li, Controllability and observability of CPSs under networked adversarial attacks. IET Control Theory Appl. **11**(10), 1596–1602 (2017)
16. Q. Hu, D. Fooladivanda, Y.H. Chang, C.J. Tomlin, Secure state estimation and control for cyber security of the nonlinear power systems. IEEE Trans. Control Netw. Syst. (2017)
17. A. Sargolzaei, K.K. Yen, M.N. Abdelghani, Preventing time-delay switch attack on load frequency control in distributed power systems. IEEE Trans. Smart Grid **7**(2), 1176–1185 (2016)
18. A. Sargolzaei, K. Yen, M. Abdelghani, Delayed inputs attack on load frequency control in smart grid, in *2014 IEEE PES Innovative Smart Grid Technologies Conference (ISGT)* (IEEE, 2014), pp. 1–5
19. A. Sargolzaei, K.K. Yen, M. Abdelghani, Time-delay switch attack on load frequency control in smart grid. Adv. Commun. Technol. **5**, 55–64 (2013)
20. A. Sargolzaei, K.K. Yen, M.N. Abdelghani, A. Mehbodniya, S. Sargolzaei, A novel technique for detection of time delay switch attack on load frequency control. Intell. Control Autom. **6**(04), 205 (2015)

21. A. Abbaspour, P. Aboutalebi, K.K. Yen, A. Sargolzaei, Neural adaptive observer-based sensor and actuator fault detection in nonlinear systems: application in UAV. ISA Trans. **67**, 317–329 (2017)
22. A. Abbaspour, K.K. Yen, S. Noei, A. Sargolzaei, Detection of fault data injection attack on UAV using adaptive neural network. Procedia Comput. Sci. **95**, 193–200 (2016)
23. M.M. Rana, Attack resilient wireless sensor networks for smart electric vehicles. IEEE Scns. Lett. **1**(2), 1–4 (2017)
24. S. Noei, A. Sargolzaei, A. Abbaspour, K. Yen, A decision support system for improving resiliency of cooperative adaptive cruise control systems. Procedia Comput. Sci. **95**, 489–496 (2016)
25. M. Ghanavati, A. Chakravarthy, P. Menon, Pde-based analysis of automotive cyber-attacks on highways, in *American Control Conference (ACC), 2017* (IEEE, 2017), pp. 1833–1838
26. P. Quinn-Judge, Cracks in the system. TIME Mag. (2002)
27. J. Leyden, Polish teen derails tram after hacking train network. *Register* **11** (2008)
28. P. Boulos, A. Sargolzaei, A. Ziaei, S. Sargolzaei, Pacemakers: a survey on development history, cyber-security threats and countermeasures. Int. J. Innov. Stud. Sci. Eng. Technol. **2**(8) (2016)
29. A. Sargolzaei, K. Yen, M. Abdelghani, A. Abbaspour, S. Sargolzaei, Generalized attack model for networked control systems, evaluation of control methods. Intell. Control Autom. **8**(3), 164–174 (2017)
30. A. Sargolzaei, K.K. Yen, M. Abdelghani, Control of nonlinear heartbeat models under time-delay-switched feedback using emotional learning control. Int. J. Recent Trends Eng. Technol. **10**(2), 85 (2014)
31. G. O'Brien, S. Edwards, K. Littlefield, N. McNab, S. Wang, K. Zheng, Securing wireless infusion pumps (NIST Special Publication, 1800), p. 8B
32. B. Satchidanandan, P. Kumar, Dynamic watermarking: active defense of networked cyber-physical systems. *Proc. IEEE* **105**(2), 219–240 (2017)
33. M. Pajic, I. Lee, G.J. Pappas, Attack-resilient state estimation for noisy dynamical systems. IEEE Trans. Control Netw. Syst. **4**(1), 82–92 (2017)
34. D. Ding, G. Wei, S. Zhang, Y. Liu, F.E. Alsaadi, On scheduling of deception attacks for discrete-time networked systems equipped with attack detectors. Neurocomputing **219**, 99–106 (2017)
35. D. Ding, Z. Wang, D.W. Ho, G. Wei, Observer-based event-triggering consensus control for multiagent systems with lossy sensors and cyber-attacks. IEEE Trans. Cybern. (2016)
36. F.-Y. Wang, D. Liu, Networked control systems, in *Theory and Applications* (Springer, London, 2008)
37. A. Sargolzaei, Time-delay switch attack on networked control systems, effects and countermeasures (2015)
38. Y.W. Law, T. Alpcan, M. Palaniswami, Security games for risk minimization in automatic generation control. IEEE Trans. Power Syst. **30**(1), 223–232 (2015)
39. J. Weiss, Industrial control system (ICS) cyber security for water and wastewater systems, in *Securing Water and Wastewater Systems* (Springer, 2014), pp. 87–105
40. A. Cardenas, S. Amin, B. Sinopoli, A. Giani, A. Perrig, S. Sastry, Challenges for securing cyber physical systems, in *Workshop on Future Directions in Cyber-physical Systems Security*, vol. 5 (2009)
41. J. Meserve, Sources: staged cyber attack reveals vulnerability in power grid, *CNN.com*, vol. 26, 2007
42. A. Greenberg, Hackers cut cities power. Forbes (Jaunuary, 2008)
43. E. Byres, J. Lowe, The myths and facts behind cyber security risks for industrial control systems, in *Proceedings of the VDE Kongress*, vol. 116 (2004), pp. 213–218
44. A.A. Cárdenas, S. Amin, S. Sastry, Research challenges for the security of control systems, in *HotSec* (2008)
45. Y. Liu, P. Ning, M.K. Reiter, False data injection attacks against state estimation in electric power grids. ACM Trans. Inf. Syst. Secur. (TISSEC) **14**(1), 13 (2011)

46. N. Rashid, J. Wan, G. Quirós, A. Canedo, M.A. Al Faruque, Modeling and simulation of cyber-attacks for resilient cyber-physical systems
47. J. Wan, A. Canedo, M.A. Al Faruque, Security-aware functional modeling of cyber-physical systems, in *2015 IEEE 20th Conference on Emerging Technologies & Factory Automation (ETFA)* (IEEE, 2015), pp. 1–4
48. H.C. Chen, M.A.A. Faruque, P.H. Chou, Security and privacy challenges in iot-based machine-to-machine collaborative scenarios, in *Proceedings of the Eleventh IEEE/ACM/IFIP International Conference on Hardware/Software Codesign and System Synthesis* (ACM, 2016), p. 30
49. A. Sargolzaei, C.D. Crane, A. Abbaspour, S. Noei, A machine learning approach for fault detection in vehicular cyber-physical systems, in *2016 15th IEEE International Conference on Machine Learning and Applications (ICMLA)* (IEEE, 2016), pp. 636–640
50. A. Sargolzaei, K.K. Yen, M. Abdelghani, S. Sargolzaei, B. Carbunar, Resilient design of networked control systems under time delay switch attacks, application in smart grid. IEEE Access (2017)
51. M. Ma, H. Chen, X. Liu, F. Allgöwer, Distributed model predictive load frequency control of multi-area interconnected power system. Int. J. Electr. Power Energy Syst. **62**, 289–298 (2014)
52. C.-T. Chen, *Linear System Theory and Design* (Oxford University Press, Inc., 1995)

Chapter 7
Detecting Community Structure in Dynamic Social Networks Using the Concept of Leadership

Saeed Haji Seyed Javadi, Pedram Gharani and Shahram Khadivi

7.1 Introduction

The concept of sustainable interdependent networks covers the interactions among a wide range of networks, including transportation networks, energy systems, communication networks, social networks, and wireless sensor networks. Online marketing via twitter [1], load management in smart power grids [2], and energy-saving in wireless networks [3] are three examples that demonstrate the interdependency of such networks. The advent and growing popularity of online social networks such as Facebook, LinkedIn, and Twitter had a significant impact on the study of social networks. One of the most critical topics in the social network analysis is the problem of finding latent communities. A community in a network is a set of nodes that are densely interconnected to one another while loosely connected to the rest of the network [4]. One could analyze and understand the structures and functions of a complex network profoundly by detecting the communities. This principal problem has been studied heavily in the past decade. A large number of fast and accurate methods have been developed by researchers in various fields of study [5]. One of the premier measurement functions called *modularity* was introduced by Newman et al. to evaluate the quality of detected structures in communities [6]. The concept of modularity let a new category of methods emerged to detect densely connected nodes in complex networks [7–9]. Their strategy was to find a decent clustering by maximizing the modularity function. Since maximizing the modularity pertains to the class of NP-complete problems [10], several heuristic approaches were proposed to find the near-optimal

S. H. S. Javadi · P. Gharani (✉)
School of Computing and Information, University of Pittsburgh, Pittsburgh, USA
e-mail: peg25@pitt.edu

S. Khadivi
eBay Inc., Aachen, Germany

© Springer International Publishing AG, part of Springer Nature 2018
M. H. Amini et al. (eds.), *Sustainable Interdependent Networks*,
Studies in Systems, Decision and Control 145,
https://doi.org/10.1007/978-3-319-74412-4_7

community structure [8, 11]. Finding the local community of a given node is another strategy in community mining which has shown its efficiency in the face of very large complex networks like World Wide Web [5]. The community is usually formed by expanding from an initial "seed" node as long as the defined local metric strictly improves [12–14]. Most of the seed-centric community detection solutions are sensitive to the position of initial source nodes as forming a local cluster around a low-degree node usually results in poor quality.

There have been extensive works focusing on finding communities in static networks [5]; however, until recently, most of the proposed algorithms ignore the dynamic aspects of social networks by discarding time information of interactions. In the static approach, a social network is treated as a single constant graph that is mostly derived by aggregation of the whole network over time. However, in reality, due to the dynamic nature of the social networks, they continuously evolve. These changes could be joining (leaving) actors to (from) the network and establishing new connections or destroying the existing ones [15]. This makes a highly dynamic network which witnesses a wide variety of changes. Examples include online social networks such as Facebook [16], email exchange networks [17], blogosphere [18]. Since these methods ignore dynamic information associated with ever-changing social networks, they can neither capture the evolutionary behavior of the network nor predict future status of community structure.

Over recent years, there has been a new trend in devising efficient algorithms to detect communities as well as tracking them in dynamic social networks. A dynamic network can be modeled as a sequence of static networks called graph snapshots where each snapshot corresponds to a particular timestep.

As we will discuss in more detail in related work, some methods employ a two-stage mining approach in which at first, a static clustering method is applied on all snapshots, and then, obtained communities will be compared with one another to track the evolution of community structure over time. Since computing communities is usually independent of the history, detected community structure of every certain snapshot is dramatically different from the ones related to the other snapshots, especially in noisy datasets. Another type of method attempts to find a good clustering for each snapshot, whereas the detected partition is not different from its history. This could be done by optimizing an objective function composed of two variables named *community quality* and *community history*. The main drawback of this approach is that it is not parameter-free. For instance, the method proposed in [19] requires the number of desired communities which is usually unknown in practice.

A community, considered as an evolving structure, can be detected by an incremental updating. The strategy is to keep the community structures of previous steps unless any changes occur in the underlying network, that is to say, whenever a new link is added or an existing one is removed, an update procedure is applied to adapt the clustering to the new structure. It has been demonstrated that this strategy is fast and it keeps community smoothness over time [20, 21]. Yet, highly changing networks could compromise the quality of the result of an incremental community detection algorithm.

The problem of finding communities in an evolving network is addressed in the literature of multiplex network analysis as well. A multiplex network is defined as a set of networks linked through interconnected layers. Each layer is composed of the same set of nodes which may be interconnected with different types of links. Multilayer networks can also be used to model dynamic networks. In [22], authors introduced a multi-slice generalization of modularity measure to quantify the quality of the detected community structure in a multiplex network.

Apart from being dynamic, another significant characteristic of real-world social networks is the presence of leaders [1], i.e., influential members in local communities, which is an old topic in the field of social science [23]. Until now, finding leaders and analyzing their social influences in various types of social networks is still an appealing topic [21, 25, 26]. Regarding this important property, recently there has been a new trend of community detection methods using community leaders as the pivotal members. The central position of a leader makes it a good option to be chosen as the initial source node in the seed-centric local community detection methods. Hence, different definitions of centrality were applied to distinguish desirable leaders from non-leader members [27].

Following this idea, we designed an efficient method to incrementally detect the communities in the dynamic social networks using the intuitive idea of importance and persistence of community leaders over time. Briefly, in the proposed method, community leaders of each timestep are detected efficiently, and the community membership of next timestep is determined by using these influential members as the initial seed nodes. Our proposed method has the following fundamental properties:

- **Online**: In order to find the meaningful communities at each timestep, the structural information of the previous snapshot is employed.
- **Fast**: Since our algorithm employed an incremental fashion to the evolving communities, it does not need to calculate the communities of each snapshot from scratch, and the runtime of the method is reduced considerably.
- **Smooth**: In real-world social networks, community memberships are not expected to change abruptly, so the detected community structure of adjacent graph snapshots should not differ a lot.
- **Parameter-free**: Unlike most of the solutions proposed in the literature, our algorithm does not require any parameter settings.
- **Adaptive**: Unlike most incremental methods, this algorithm is efficient even for highly dynamic networks.

The remainder of this paper is organized as follows: In the following section, we review the existing approaches related to our work. In Sect. 7.3 after the preliminaries, the problem definition is provided. Then, we present the proposed method in more details. Section 7.4 is dedicated to evaluating the technique by extensive experiments on several real-world and synthetic dynamic networks. Finally, we present conclusions and future directions of our research in Sect. 7.5.

7.2 Related Work

In this section, a survey on some existing works related to the contributions of the proposed method is provided. First, we review several notable seed-centric methods, and then, we discuss three main existing approaches to detect the community structure in the dynamic social networks.

7.2.1 Seed-Centric Approach

In addition to a vast number of designed algorithms to unfold the community structure of a complex network by maximizing a global fitness function like modularity, an alternative approach is introduced recently to detect the communities. The main idea of this approach, which is called seed-centric, is to determine local community around a given node. At first, it was accepted as a suitable approach for finding local community structure; subsequently, several local quality functions were introduced in [12–14] to measure the goodness of the detected local clusters. The authors also devised greedy methods to expand a local community around the source node by maximizing their quality measure. Since these methods are sensitive to the position of initial seed nodes, Chen et al. [28] chose the local maximal degree as the seed of their algorithm and formed the local cluster by iteratively adding neighbor nodes to it. To add suitable adjacent nodes to the core of the local communities, they have tested multiple quality functions. Although they reached higher accuracy compared with the other local community detection methods, due to the high time complexity of their approach, it is not scalable in finding communities in large social networks.

Seed-centric approach is not limited to the local community detection problem; DOCNet is another efficient algorithm to identify overlapping communities on the entire network. The authors used two concepts of *compactness* and *separability* [29] to create a new objective function called *index of connectivity*. Their main strategy is to select a central influential node and add proper nodes to expand the community until a stopping criterion is met [30]. Since in these works the initial seed nodes are high central members, they are usually referred as community core or community leaders [27]. With the assumption that every group of individuals in communities is composed of two types of members, leaders, and followers, Rabbany et al. [31] found the k-most central nodes as top leaders which could be followed by non-leader nodes to form communities. They examined a wide variety of central measures to find the most appropriate definition of leaders. However, their method is not parameter-free and requires the number of the desired communities, which most of the time is unknown in advance. In [32], the characteristics of major seed-centric methods are discussed in more depth.

7.2.2   Community Detection in Dynamic Networks

Usually, dynamic networks are represented as a series of static graphs over a period of discrete timesteps, called snapshots. Each snapshot corresponds to a particular timestep of the dynamic network. A community in a dynamic social network is not only a group of densely interconnected nodes, but its members keep their close interactions over an expected time. This problem is also known as dynamic graph clustering. Most of the proposed methods to detect communities in temporal networks could be categorized into three main strategies:

The first strategy is a *two-stage* method (also known as *independent clustering*) which is based on slicing the whole network as a series of snapshots. Its basic idea is to apply a static clustering method to each snapshot independently and captures the evolution of the communities by comparing the clustering of the consecutive snapshots. Based on this approach, Hopcroft et al. [33] were of the first authors who applied a static clustering on each snapshot, and they tracked communities over time by the clusters' similarities over consecutive timesteps. They showed that even small perturbations in the graph could lead significant changes in the structure of the detected communities. This is the main disadvantage of two-stage methods; they are vulnerable to even small changes in network structures, and so the output of these methods would result in noisy and short-life communities. To resolve this problem, the authors introduced the *natural communities* which could remain stable under several perturbations in a graph structure. Based on the same consideration and with a different methodology, Seifi and Guillaume [34] tracked the evolution of community cores instead of all members of communities. They defined community cores as a set of nodes that frequently clustered together during several executions of a nondeterministic community detection method. They showed that in a dynamic social network, this group of nodes is much more stable compared with all members of a community.

Although core-based methods could reduce the variance caused by unstable nodes, they usually require parameters to specify the borderline of core and noncore members. Moreover, applying several perturbations and applying clustering method on each snapshot to detect significant clusters are very time-consuming. The other drawback of independent clustering is the computing similarities between huge numbers of the communities across multiple snapshots. Since the number of the found communities of each snapshot could be more than the number of nodes, it could be impractical in facing big dynamic networks like the method introduced in [35].

Apart from independent clustering, *evolutionary clustering* was introduced. In this approach, by using the structural information of the preceding timesteps, the detected communities of the consecutive snapshots should not dramatically different. In doing so, these methods try to optimize an objective function composed of two variables: the cost of a good partitioning for all snapshots and the cost of the difference between the structure of the communities and its history. FacetNet is a well-known evolutionary framework to find a soft community structure based on generative models. In their proposed framework, the community structure of the current snapshot is detected by incorporating the current graph structure and historic community evolution patterns [36]. Recently, Kawadia and Sreenivasan [37] have proposed a new measure called "Estrangement" to quantify the partition distance between two consecutive timesteps. They describe Estrangement as "the fraction of intra-community edges that become intercommunity edges as the network evolves to the subsequent snapshot." Applying this constraint on the distance metric, they found temporal communities in each snapshot by maximizing the modularity.

Another strategy to unfold communities in temporal networks is *incremental community detection*, which is based on structural changes of the network such as link insertion and link removal. Methods using this approach consider the evolution of networks as a series of atomic events which could change community memberships of individuals (therefore, it is known as *event-based* clustering). This strategy has two main advantages when encountering with any changes in the network structure: First, the running time of the algorithm is decreased significantly because it does not need to apply clustering method from scratch. Second, it preserves community smoothness over time because of its dynamic updating strategy.

Gorke et al. [38] introduced dGlobal which is a dynamic version of the CNM method [39] to maximize modularity at each timestep. They aimed to save and modify the dendrogram of the obtained clusters. Community memberships of affected nodes will be determined simultaneously, proportional to structural changes in the network. Similar to that, Nguyen et al. [15] proposed another algorithm to find modular communities in each snapshot. Although their method is faster than the method which clusters each snapshot independently [7], running their method for a long time on a dynamic network will end up in poor quality results. The method proposed by Takaffoli et al. [40] intends to find the community structure at any time based on the extracted clusters from the previous timestep. They introduced an adaptive algorithm which like two-stage methods employs an event-tracking framework. At each timestep, they consider the obtained connected components of the communities in the last snapshot as the initial seeds for the current snapshot. They form communities around the seeds by maximizing the ratio of the average internal degree and minimizing the average external degree of the local cluster. However, they did not describe how their framework handles some certain scenarios. An example is a case in which new emerged nodes do not belong to any existing communities and independently create a new community.

Event-based clustering implicitly assumes that network connections do not change much over time, and even so, they have an only small impact on the current community structures. But, in contrast, most of the real-world social networks do change dramatically even in a short period [41, 42]. In this regard, Barabasi stated that dynamics of many social and economic phenomena are driven by bursty nature of human behavior [43]. Speaking of dynamic social networks, Wang et al. [44] through their experiments showed that a large number of nodes appear in less than tree snapshots. In other words, only a small portion of nodes could remain stable during network changes over time. Recently, Xu et al. [45] studied the problem of detecting stable community cores in mobile social networks. By assuming that the links existing in the community core remain stable during consecutive snapshots, they summarized the problem of detecting dynamic community structure in analyzing community core evolution over time.

7.3 Method Description

7.3.1 The Main Idea

A good community structure in a dynamic social network is one in which the members should have a strong structured similarity with each other, and they should keep this similarity over time. As a result, any method whose aim is to detect and track communities in dynamic social networks should consider two main characteristics:

1. Detected communities should be modular at each timestep. In other words, nodes tightly connected to one another have to be grouped in the same cluster.
2. The two-temporal smoothness of clusters over consecutive snapshots should be preserved; that is, in most of the cases, the communities of time t should not sharply differ from the ones of time t−1.

In this section, we first present some definitions focusing on the terms that we use in this paper. Then, we explain the existence and importance of community leaders in real-world social networks. The formal definition of a leader is introduced just after the significant role of leaders is expressed. The next section is dedicated to introducing our local clustering method expanding communities around the promising leaders. In the following, we discuss an important phenomenon in dynamic social networks, i.e., the advent of new communities. Finally, our robust incremental model will be described.

7.3.2 Notations and Definitions

We model a dynamic social network as a sequence of snapshots $\{G^1, G^2, \ldots, G^\Delta\}$, where $G^t = (V^t, E^t)$ denotes a static graph at time point $1 \leq t < \Delta$, V^t and E^t are the set of nodes and edges of graph G^t, respectively. The degree of node v^t is denoted by d_{v^t} which is the number of neighbors of v^t at time point t. The in-community degree of v^t is given by $d_{v^t}^{in}$ which indicates the number of links that connect v^t to nodes of the same cluster.

We use $L^{t-1} = \left\{L_1^{t-1}, L_2^{t-1}, \ldots, L_j^{t-1}\right\}$ to represent the set of leaders of j communities in the snapshot of G^{t-1}, and $C^t = \left\{C_1^t, C_2^t, \ldots, C_j^t\right\}$ denotes the set of communities that are obtained by expanding around corresponding leaders of the previous snapshot, i.e., L^{t-1}. Since some nodes may be remained unassigned to any clusters, the set $P^t = \left\{ C_1^t, C_2^t, \ldots, C_j^t, C_{j+1}^t, \ldots, C_k^t \right\}$ is introduced. The set of newcomer communities $\left\{ C_{j+1}^t, \ldots, C_k^t \right\}$ is a subset of P^t which is determined by clustering the remaining nodes. Consequently, the set $P^t = \left\{ C_1^t, C_2^t, \ldots, C_k^t \right\}$ partitions G^t into some nonoverlapping communities where $\bigcup_{1 \leq i \leq k} C_i^t = G$ and $C_i \cap C_j = \varnothing \, \forall i, j$.

7.3.3 Existence of Leaders in Social Networks

Social networks reflect the behavior and interactions of individuals; and as in real-world, some members have great influence on the others. Communities are commonly formed around these influential members called leaders. This observation is one of the results of "preferential attachment" model introduced by Barabasi [46] to explain the power-law distributions of the node degrees in the real-world social networks. In their model, a newcomer node preferentially connects to the nodes that have more connections. These high degree nodes potentially are more popular and have stronger ability to attract new members. This mechanism reproduces the power-law degree distribution observed in the real social networks. When the degree distribution of a graph follows power laws, it means that a major fraction of the nodes are low-degree and conversely, a small number of them have so many connections (Fig. 7.1). This key property has been studied in many real-world networks such as World Wide Web [47], citation network [48], and online social networks [49].

Fig. 7.1 A random network generated by the model of Barabasi and Albert

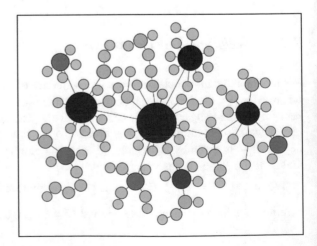

Based on the discussion above, a node with a high degree centrality has a high attraction to absorb other nodes and hence to form a community. This is one of the main characteristics of leaders which could be captured from the local structure of members in a social network. Based on this important feature, it is argued recently for choosing the most central node as the initial seed to detect meaningful local communities. Among dozens of centrality parameters, the maximal-degree nodes are showed to be good choices to select in seed-centric community detection algorithms [30, 31, 50].

The topology of initial seeds is one of the most important criteria in seed-centric community detection methods. Many of the existing methods expand the community around a single node [31, 50], whereas choosing a set of central nodes as the initial seeds lead to fewer computations and may result in a more robust solution. On the other hand, to ensure obtaining high-quality communities, the set of initial nodes should be strictly interconnected to one another.

7.3.4   Detecting Community Leaders

In the proposed method, local communities of each snapshot are determined by initialized seed nodes. So, defining and selecting the leaders have a significant effect on detecting desired clusters. A set of nodes with particular characteristics will be selected as the promising community leaders. These nodes are densely interconnected and form a clique structure. Since a clique is a subgraph whose underlying nodes are fully connected, we could be sure that they would be clustered together when any community detection method is applied. Also, in [5, 51] it is showed that it is better to choose a clique instead of an individual node as an initial seed.

Suppose community C_i^t is known, we present the definition of community leader of C_i^t as follows:

Community leaders: Considering node v has the maximum in-community degree of C_i^t, community leaders of this cluster are defined as the intersect of nodes that exist in all maximal cliques containing v.

A maximal clique is a fully connected subgraph which is not a subset of any other cliques. To find maximal cliques in a network, we use the method proposed by Eppstein et al. [52] which is fast and efficient. Because the method is applied only on the ego network of the node with a maximum in-community degree and not on the whole network, we can assure that the total running time of leader detection process would be short.

In dynamic social networks, we expect the leaders not only to construct the communities but to be more stable than non-leader nodes over time also. Experimental results on real-world social networks strictly confirm this claim.

7.3.5   Local Expansion

To associate follower nodes to their correspond leaders, we use "index of connectivity" which is an unbiased local objective function presented by [30], and it is defined as follows:

$$IC_{C_i} = \frac{\eta_{C_i} - \mu_{C_i}}{\sqrt{\eta_{C_i} + \mu_{C_i}}} \tag{7.1}$$

where η_{C_i} represents the number of links connecting members of community C_i^t to one another, and μ_{C_i} is the number of links connecting members of community C_i^t to the outside of the community. This objective function aims to strengthen internal edges as well as weaken external edges of the cluster. To expand community around the leaders, nodes will be selected that greedily increase the index of connectivity. To recalculate the objective function for each candidate node efficiently, we may use the following equation:

$$I\acute{C}_{c_i^t} = \frac{\eta'_{c_i^t} - \mu'_{c_i^t}}{\sqrt{\eta'_{c_i^t} + \mu'_{c_i^t}}} \tag{7.2}$$

where $\eta'_{C_i^t}$, $\mu'_{C_i^t}$, and $I\acute{C}_{C_i^t}$ are corresponding scores after aggregating the node u into C_i^t. Two variables $\eta'_{c_i^t}$ and $\mu'_{c_i^t}$ are determined as follows:

$$\eta'_{C_i^t} = \eta_{C_i^t} + in_{C_i^t}(u) \tag{7.3}$$

$$\mu'_{C_i^t} = \mu_{C_i^t} + d(u) - 2in_{C_i^t}(u) \tag{7.4}$$

where $d(u)$ and $in_{C_i^t}(u)$ are the degree and in-community degree of node u. We continuously keep expanding the community C_i^t until no further improvement on the objective function could be achieved.

7.3.6   Membership of Nodes

After all communities are expanded, we would face three type of nodes: (1) Nodes that are assigned to one community, (2) Nodes that are assigned to more than one community and usually sit on the intersections of communities, and (3) Nodes that are not assigned to any community because they show low orientation to any existing leader.

In real-world social networks, individuals are often shared between communities [5]. However, to simplify the model, the most suitable community of each hub will be determined. This is based on the ratio of the intersection of neighbors of u and members of C_i^t to the union of them all. Hub u would merge to the community with the highest similarity. The similarity measure is calculated as follows:

$$J\left(u, C_i^t\right) = \frac{\left|neighbors(u) \cap members\left(C_i^t\right)\right|}{\left|neighbors(u) \cup members(C_i^t)\right|} \tag{7.5}$$

In case 3 where nodes are not associated with any leader, they may be considered to have their community. In the next subsection, we will discuss this common event happening in dynamic social networks.

7.3.7   Advent of New Communities

Through our experiments, we noted that a considerable number of nodes would appear at each new timestep. Usually, the newcomer members are interconnected dense enough to form a new community. Therefore, we expect them not to be associated with any existing community and to remain unassigned. In this case, identified communities are temporarily omitted from the network, and a static method is applied to cluster the unassigned nodes. Newly detected clusters are considered as newborn communities. This is an important event that most other incremental methods have failed to handle it properly.

7.3.8   Algorithm Description

Our approach requires an initial set of leaders L^1 for the first snapshot. Thus, any static community detection algorithm would be applied on the first snapshot of the network to identify communities of graph G^1. With obtained clusters C^1, leaders of corresponding communities could be determined easily by definition proposed in Sect. 7.3.4. We represent leaders of the first snapshot by L^1. The leaders are the most cohesive structures located around the node of highest in-community degree. Featuring the durability of leaders over time, they could be suitable representatives of the community for the next timestep. By facing a new snapshot, the set of leaders of the current timestep will be employed as the initial seeds for next stage. New communities will locally expand around the seed nodes by employing the procedure discussed in Sect. 7.3.5.

After local communities are expanded, we choose the most suitable communities for the hub nodes. In addition, it is very likely to have some nodes that may belong to no cluster. In other words, identified cluster $C^t = \left\{ C_1^t, C_2^t, \ldots, C_j^t \right\}$ which are formed around the $L^{t-1} = \left\{ L_1^{t-1}, L_2^{t-1}, \ldots, L_j^{t-1} \right\}$ do not partition graph G^t. In this case, identified communities are omitted, and a static clustering method like Infomap [53] is used to cluster the remained nodes. We use $\acute{C}^t = \left\{ C_{j+1}^t, C_{j+2}^t, \ldots, C_k^t \right\}$ to denote new clusters detected by the static clustering method. By union of C^t and $\acute{C}^t$ we partition graph G^t as $P^t = C^t \cup \acute{C}^t$. We note that all communities obtained in the second step are considered as newborn communities. Similarly, a community dissolves when all of its corresponding leaders disappear in the next snapshot.

7.4 Experiments

In order to evaluate the performance of the proposed method, we conducted comprehensive experiments on both computer-generated networks and real-world dynamic ones. Two networks generated by computer and also several real-world dynamic networks were investigated. Studying the persistence of the leaders and non-leader nodes on real-world social network proves that the community leaders of the proposed definition are much more stable in comparison with follower nodes. As a baseline for comparison, we applied three well-known community mining algorithms to different approaches. These methods are FacetNet [19], Estrangement [37], and static clustering using FUA method [7]. The static method does not consider any temporal evolution and is applied on each snapshot independently. We consider running time, community smoothness, and similarity of the result to ground-truth communities as the evaluation criteria. Before the experiments, both real-world and synthetic datasets are first described.

7.4.1 Real-World Dynamic Social Networks

We tested our method on different real-world dynamic social networks selected from various fields of studies such as online social networks, email, and cell phone communication networks.

Catalano cell phone network [54] consists of communication information of 400 unique telephone numbers for 10 months. The total number of phone calls during this period is 9834. Each corresponding snapshot is created by aggregating all connections during every 24 h.

Opsahl network originates from an online community for students at the University of California, Irvine [55]. It consists of 59835 messages sent over 1899 subscribed users from April to October of 2004. Collected information of every two weeks represents a single timestep in our experiments.

Another network is composed of a set of exchanged emails of the department of computer science at KIT [55]. This network was collected over four years, and it is comprised of 1890 users and about 550000 email messages. The corresponding chair ID of each email address is considered as ground-truth community structure. We have sliced this data into snapshots of 1 week for our experiments.

Another network contains the wall posts from the Facebook New Orleans networks [56]. In this dataset, each edge represents a post to other user's wall, and each node represents a Facebook user. This dynamic network was gathered in about 30 months containing 46952 nodes and 876993 links. The time interval between snapshots is one month.

The Enron email network [57] consists of about 1100000 emails which are sent among 87273 employees of Enron Corporation between the beginning of 2001 and 2003, whose graph snapshots are taken monthly.

7.4.2   Computer-Generated Networks

To investigate the accuracy of the proposed algorithm, we tested our method on two artificially generated dynamic networks. In these dynamic graphs, hidden ground-truth communities are embedded in the network data and evolve randomly over time to simulate the evolution of real-world communities.

Kawadia and Sreenivasan [37] introduced a dynamic network generator based on Markovian evolution. In their benchmark, a set of consecutive snapshots is generated consisting predefined communities embedded in a random background. The underlying links in the communities undergo Markovian evolution to simulate community evolution in real-world scenarios. The initial snapshot is created by Erdos–Renyi random graph model [58] in which N represents the number of nodes and p_r is the existence probability of an edge in the random background. Parameter p indicates the probability of intra-community edges which is independent of p_c. An edge disappears from the community by the probability of p, and conversely, a new link appears by the probability q. To preserve initial edge density within the community, q is set as $q = \frac{pp_c}{(1-p_c)}$.

Considering each community characterized by a series of evolutionary events, Green et al. [59] proposed synthetic dynamic networks based on LFR benchmark [60]. They created four networks corresponding to four main events occurred during the lifetime of a community. These events are as follows:

Intermittent communities: In the corresponding network, 10% of communities are removed randomly at each timestep.
Expansion and contractions: To examine the effect of rapid expansions and contractions of communities, 40 randomly selected communities to absorb new nodes or lose their former members by 25% of the previous size.
Birth and death: At each timestep, 40 new clusters are created by the nodes which have left their former communities. Furthermore, 40 existing communities are removed randomly.

Merging and splitting: In the last case, 40 instances of the existing clusters merged two by two, and similarly, 40 communities split into two new communities at each snapshot.

7.4.3 Persistence of Leaders in Real-World Networks

Before we compare the performance of our algorithm with the baseline methods, we examine the stability of community leaders over time. It is illustrated in Fig. 7.2 that the persistence of leaders versus community followers on all timesteps in social

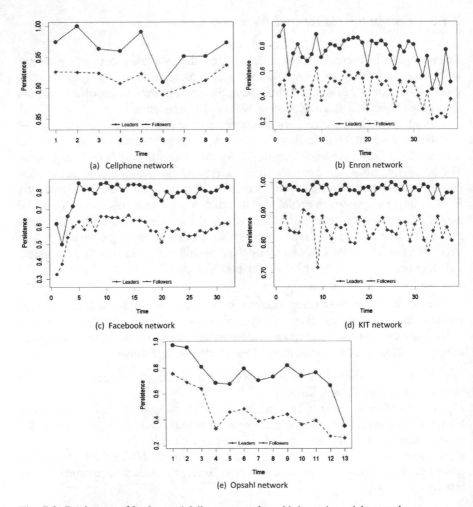

Fig. 7.2 Persistence of leaders and follower on real-world dynamic social networks

networks is introduced. The vertical line of time t shows the ratio of community leaders/followers presented on both G^t and G^{t+1}. As it can be seen, the persistence of community leaders is always significantly greater than that of follower nodes. For instance, in time step 10 in the KIT network, about 30% of followers are missed between G^{10} and G^{11}, while all community leaders are present on both timesteps.

Note that community leaders for each timestep are determined based on the community structure of that snapshot. In other words, the role of community leadership could shift to other members of the group.

7.4.4 Temporal Smoothness

In the next experiment, we analyze the smoothness of identified communities. Therefore, clustering of successive snapshots is compared together by a similarity measure. Higher similarity value indicates a smoother evolving community structure. To compare the community memberships of two adjacent snapshots, normalized mutual information (NMI) is adopted. Let A and B be two partitions of the same graph, NMI(A, B) equals 1 if A and B are identical and equals 0 if they are totally different. The greater the value of NMI, the more similar the two partitions are.

In our datasets, some nodes are removed, and new members are added over time. As a result, the size of two consecutive snapshots may differ, and so NMI cannot be applied. To overcome this issue, we measure the clustering similarity of members attended in both consecutive snapshots.

We should note that since FacetNet requires an adjacency matrix for each snapshot to calculate community membership of nodes, its space complexity is more than $\Omega(\Delta n^2)$. That is, FacetNet is impractical for networks more than ten thousand nodes. Also, the execution time of Estrangement method is very high, as it does not converge to any solution in 24 h of runtime. That is why we have omitted the solutions of these two methods for Facebook and Enron datasets.

As we can see in Figs. 7.3 and 7.4, our method achieves higher temporal smoothness compared with several baseline methods, except for Opsahl and cell phone datasets in which the proposed algorithm and FacetNet have a close competition. As expected, the independent clustering method ranks last because it does not use any structural information of previous partitions to cluster the current graph snapshot.

7.4.5 Similarity to Ground-Truth Communities

In the next evaluation phase, we investigate the similarity of clusterings to the ground-truth communities. In order to measure the similarity, we employed NMI as

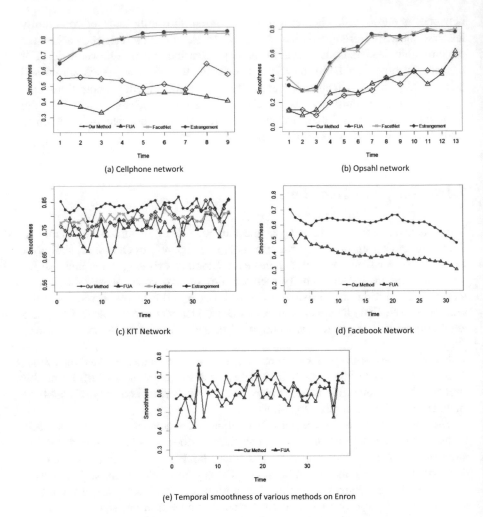

Fig. 7.3 Temporal smoothness of dynamic clustering methods on real-world social network

well. We compare the performance of proposed method with the other approaches on KIT. Among the available datasets, KIT is the only real-world social network which contains metadata as the ground-truth community structure.

The results in Fig. 7.5 confirm that detected communities of the proposed method are the most similar partitions to the known community structure of KIT. Although the FacetNet is the next accurate method, it requires the number of desired communities as input which could result in completely different solutions using different parameter settings. In order to ensure a fair comparison, this parameter is set as the average number of known communities in KIT. However, in practice, the true number of clusters is unknown beforehand.

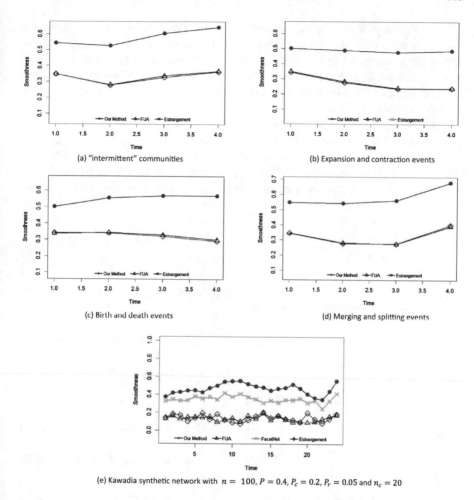

(a) "intermittent" communities

(b) Expansion and contraction events

(c) Birth and death events

(d) Merging and splitting events

(e) Kawadia synthetic network with $n = 100, P = 0.4, P_c = 0.2, P_r = 0.05$ and $n_c = 20$

Fig. 7.4 Temporal smoothness of dynamic clustering methods on synthetic datasets

Fig. 7.5 Similarity to ground-truth community in KIT network

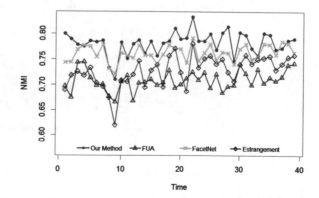

The accuracy of Estrangement method considerably fluctuates over time. But overall, it shows a better performance in comparison with FUA method. A possible explanation would be that Estrangement optimizes modularity function by considering the community structure of the previous snapshot, whereas independent clustering aims at maximizing modularity merely which may cause missing some important evolution details.

After KIT is investigated, the accuracy of our algorithm is tested on both sets of computer-generated datasets (Fig. 7.6). The datasets provided by Green et al. aimed to capture four main events occurred during the lifetime of a community. The results state that the accuracy of the proposed method increases, while the performance of other algorithms remains almost constant. For Kawadia synthetic

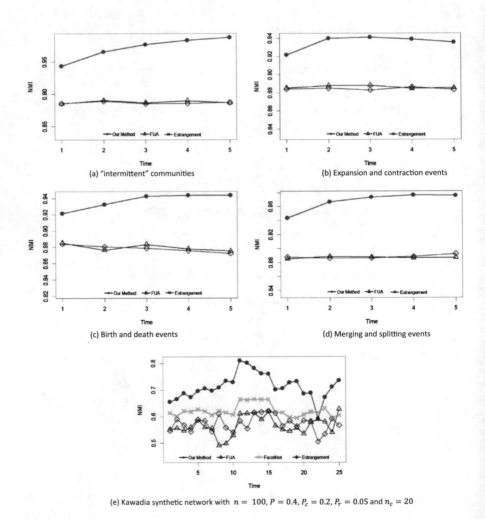

(a) "intermittent" communities

(b) Expansion and contraction events

(c) Birth and death events

(d) Merging and splitting events

(e) Kawadia synthetic network with $n = 100$, $P = 0.4$, $P_c = 0.2$, $P_r = 0.05$ and $n_c = 20$

Fig. 7.6 Similarity to ground-truth community in synthetic networks

Table 7.1 Running time of different dynamic clustering methods

	Cell phone (s)	Opsahl (s)	KIT (s)	Facebook	Enron
FacetNet	8	72	163	–	–
Estrangement	543	612	1532	–	–
Proposed method	33	14	210	32 min	36 min

network, our detected community structure is the most similar to the ground-truth community structure as well.

7.4.6   Comparing the Runtimes

In this section, we will compare the running times of dynamic clustering methods. The results are listed in Table 7.1.

The results of Facebook and Enron datasets on which Estrangement method was applied did not converge in 24 h of runtime; thus, they are omitted in Table 7.1. Moreover, the runtime results of these two datasets are ignored for FacetNet method since the corresponding process required more memory than the amount available in the test machine. It is obvious that the proposed method will be more scalable than the existing work, in confronting with big dynamic social networks.

7.5   Conclusion

Detection of communities and tracking their evolution in dynamic social networks is a significant research problem. In this work, we proposed a fast parameter-free method to find meaningful communities in highly dynamic social networks. By the concept of leadership, our algorithm first identifies some influential nodes called leaders and then employs these central members to expand local communities around them. Thanks to the incremental approach, our method reveals smooth community structure over time. We have conducted comprehensive experiments on both computer-generated networks and real-world dynamic ones. Since our method employs the community structure of the previous snapshots, it seizes the community evolution over time. The obtained results distinguish our algorithm from the existing dynamic clustering methods.

The behavior of the community leaders is of great importance in social networks. For future works, we plan to investigate the lifetime evolution of these promising members in various dynamic social networks. This information will be used to capture the main characteristics of such networks and enables us to predict the incoming structure of networks.

References

1. G. Liu, et al., Best practices for online marketing in Twitter: an experimental study. Electro Information Technology (EIT), 2016 IEEE International Conference on. IEEE, 2016
2. M.H. Amini, B. Nabi, M.-R. Haghifam, Load management using multi-agent systems in smart distribution network. Power and Energy Society General Meeting (PES), 2013 IEEE. IEEE, 2013
3. A. Mohammadi, M.J. Dehghani, Spectrum allocation using fuzzy logic with optimal power in wireless network. Computer and Knowledge Engineering (ICCKE), 2014 4th International eConference on. IEEE, 2014
4. M. Girvan, M.E.J. Newman, Community structure in social and biological networks. Proc. Natl. Acad. Sci. U. S. A. **99**, 12, 7821–7826 (2002)
5. S. Fortunato, Community detection in graphs. Phys. Rep. **486**(3–5), 75–174 (2010)
6. M. Newman, M. Girvan, Finding and evaluating community structure in networks. Phys. Rev. E (2004)
7. V.D. Blondel, J.-L. Guillaume, R. Lambiotte, E. Lefebvre, Fast unfolding of communities in large networks. J. Stat. Mech. Theory Exp. **2008**(10), P10008 (2008)
8. A. Clauset, M. Newman, C. Moore, Finding community structure in very large networks. Phys. Rev. E **70**(6), 066111 (2004)
9. X. Liu, T. Murata, Advanced modularity-specialized label propagation algorithm for detecting communities in networks. Phys. A Stat. Mech. Appl. (2010)
10. U. Brandes, D. Delling, M. Gaertler, R. Gorke, M. Hoefer, Z. Nikoloski, D. Wagner, On modularity clustering. IEEE Trans. Knowl. Data Eng. **20**(2), 172–188 (2008)
11. S.H.S. Javadi, S. Khadivi, M.E. Shiri, J. Xu, An ant colony optimization method to detect communities in social networks, in *2014 IEEE/ACM International Conference on Advances in Social Networks Analysis and Mining (ASONAM 2014)* (2014), pp. 200–203
12. A. Clauset, Finding local community structure in networks. Phys. Rev. E **72**(2), 026132 (2005)
13. F. Luo, J. Wang, E. Promislow, Exploring local community structures in large networks, in *2006 IEEE/WIC/ACM International Conference on Web Intelligence (WI 2006 Main Conference Proceedings) (WI'06)* (2006), pp. 233–239
14. A. Lancichinetti, S. Fortunato, J. Kertész, Detecting the overlapping and hierarchical community structure in complex networks. New J. Phys. (2009)
15. N.P. Nguyen, T.N. Dinh, Y. Xuan, M.T. Thai, Adaptive algorithms for detecting community structure in dynamic social networks, in *2011 Proceedings of the IEEE INFOCOM*, pp. 2282–2290, Apr 2011
16. N. Ellison, C. Steinfield, C. Lampe, The benefits of Facebook 'friends:' social capital and college students' use of online social network sites. J. Comput. Commun. (2007)
17. P. Holme, J. Saramäki, Temporal networks. Phys. Rep. **519**(3), 97–125 (2012)
18. R. Kumar, J. Novak, P. Raghavan, A. Tomkins, On the bursty evolution of blogspace, in *World Wide Web* (2005)
19. Y.-R. Lin, Y. Chi, S. Zhu, H. Sundaram, B.L. Tseng, Analyzing communities and their evolutions in dynamic social networks. ACM Trans. Knowl. Discov. Data **3**(2), 1–31 (2009)
20. G. Palla, I. Derényi, I. Farkas, T. Vicsek, Uncovering the overlapping community structure of complex networks in nature and society. Nature (2005)
21. T. Hartmann, A. Kappes, D. Wagner, Clustering evolving networks. arXiv:1401.3516, 2014
22. P.J. Mucha, T. Richardson, K. Macon, M.A. Porter, J.-P. Onnela, Community structure in time-dependent, multiscale, and multiplex networks. Science **328**(5980), 876–878 (2010)
23. L.C. Freeman, T.J. Fararo, W. Bloomberg Jr., M.H. Sunshine, Locating leaders in local communities: a comparison of some alternative approaches. Am. Sociol. Rev. 791–798 (1963)
24. L. Lü, Y.-C. Zhang, C.H. Yeung, T. Zhou, Leaders in social networks, the Delicious case. PLoS One **6**(6), e21202 (2011)

25. R. Wang, W. Zhang, H. Deng, N. Wang, Q. Miao, X. Zhao, Discover community leader in social network with pageRank, in *Advances in Swarm Intelligence* (Springer, 2013), pp. 154–162
26. T.W. Valente, Social network thresholds in the diffusion of innovations. Soc. Netw. **18**(1), 69–89 (1996)
27. R. Kanawati, LICOD: leaders identification for community detection in complex networks, in *2011 IEEE Third International Conference on Privacy, Security, Risk Trust. 2011 IEEE Third International Conference on Social Computing*, pp. 577–582, Oct 2011
28. Q. Chen, T.-T. Wu, M. Fang, Detecting local community structures in complex networks based on local degree central nodes. Phys. A Stat. Mech. Appl. **392**(3), 529–537 (2013)
29. A. Lancichinetti, S. Fortunato, J. Kertész, Detecting the overlapping and hierarchical community structure in complex networks. New J. Phys. **11**(3), 033015 (2009)
30. D. Rhouma, L. Ben Romdhane, An efficient algorithm for community mining with overlap in social networks. Expert Syst. Appl. **41**(9), 4309–4321 (2014)
31. R.R. Khorasgani, J. Chen, O.R. Zaïane, Top Leaders Community Detection Approach in Information Networks, July 2010
32. R. Kanawati, Seed-Centric Approaches for Community Seed-Centric Algorithms: A Classification Study (2014), pp. 197–208
33. J. Hopcroft, O. Khan, B. Kulis, B. Selman, Tracking evolving communities in large linked networks (2004), pp. 5249–5253
34. M. Seifi, J. Paris, J. Guillaume, Community cores in evolving networks, pp. 1173–1180 (2012)
35. G. Palla, A.-L. Barabási, T. Vicsek, Quantifying social group evolution. Nature **446**(7136), 664–667 (2007)
36. Y. Lin, FacetNet: A Framework for Analyzing Communities and Their Evolutions in Dynamic Networks (2008), pp. 685–694
37. V. Kawadia, S. Sreenivasan, Sequential detection of temporal communities by estrangement confinement. Sci. Rep. **2**, 794 (2012)
38. G. Robert, P. Maillard, C. Staudt, D. Wagner, Modularity-Driven Clustering of Dynamic Graphs (2010), pp. 436–448
39. M.E.J. Newman, Finding community structure in networks using the eigenvectors of matrices. Phys. Rev. E **74**(3), 036104 (2006)
40. M. Takaffoli, R. Rabbany, O.R. Za, Incremental Local Community Identification in Dynamic Social Networks (2013), pp. 90–94
41. S.A. Myers, J. Leskovec, The bursty dynamics of the Twitter information network, in *Proceedings of the 23rd International Conference on World Wide Web—WWW '14* (2014), pp. 913–924
42. S. Gaito, M. Zignani, G.P. Rossi, A. Sala, X. Zhao, H. Zheng, B.Y. Zhao, On the bursty evolution of online social networks, in *Proceedings of the First ACM International Workshop on Hot Topics on Interdisciplinary Social Networks Research—HotSocial '12* (2012), pp. 1–8
43. A.-L. Barabási, The origin of bursts and heavy tails in human dynamics. Nature **435**(7039), 207–211 (2005)
44. Y. Wang, B. Wu, N. Du, Community Evolution of Social Network: Feature, Algorithm and Model, p. 16, Apr 2008
45. H. Xu, Z. Wang, W. Xiao, Analyzing community core evolution in mobile social networks, in *2013 International Conference on Social Computing*, pp. 154–161, Sep 2013
46. A. Barabási, Emergence of Scaling in Random Networks. Science (80-.) **286**(5439), 509–512 (1999)
47. S. Chakrabati, B. Dom, D. Gibson, J. Kleinberg, S. Kumar, P. Raghavan, S. Rajagopalan, A. Tomkins, Mining the link structure of the World Wide Web. IEEE Comput. **32**(8), 60–67 (1999)
48. S. Bilke, C. Peterson, Topological properties of citation and metabolic networks. Phys. Rev. E **64**(3), 036106 (2001)

49. R. Kumar, J. Novak, A. Tomkins, Structure and evolution of online social networks, in *Proceedings of the 12th ACM SIGKDD International Conference on Knowledge Discovery and Data Mining—KDD '06* (2006), p. 611

50. R. Kanawati, YASCA: an ensemble-based approach for community detection in complex networks, in *Computing and Combinatorics SE - 57*, ed. by Z. Cai, A. Zelikovsky, A. Bourgeois, vol. 8591 (Springer International Publishing, 2014), pp. 657–666

51. J. Xie, S. Kelley, B.K. Szymanski, Overlapping community detection in networks. ACM Comput. Surv. **45**(4), 1–35 (2013)

52. D. Eppstein, M. Löffler, D. Strash, Listing All Maximal Cliques in Sparse Graphs in Near-optimal Time, p. 13, June 2010

53. M. Rosvall, C.T. Bergstrom, Maps of random walks on complex networks reveal community structure. Proc. Natl. Acad. Sci. U. S. A. **105**(4), 1118–1123 (2008)

54. N. Eagle, A.S. Pentland, D. Lazer, Inferring friendship network structure by using mobile phone data. Proc. Natl. Acad. Sci. **106**(36), 15274–15278 (2009)

55. R. Görke, M. Holzer, Dynamic network of email communication at the Department of Informatics at Karlsruhe Institute of Technology (KIT) (2011), http://i11www.iti.uni-karlsruhe.de/en/projects/spp1307/emaildata

56. B. Viswanath, A. Mislove, M. Cha, K.P. Gummadi, On the evolution of user interaction in Facebook, in *Proceedings of the 2nd ACM Workshop on Online Social Networks* (2009), pp. 37–42

57. J. Leskovec, K.J. Lang, A. Dasgupta, M.W. Mahoney, Community structure in large networks: Natural cluster sizes and the absence of large well-defined clusters. Internet Math. **6** (1), 29–123 (2009)

58. P. ERDdS, A. R&WI, On random graphs I. Publ. Math. Debr. **6**, 290–297 (1959)

59. D. Greene, D. Doyle, P. Cunningham, Tracking the evolution of communities in dynamic social networks, in *2010 International Conference on Advances in Social Networks Analysis and Mining*, pp. 176–183, Aug 2010

60. A. Lancichinetti, S. Fortunato, F. Radicchi, Benchmark graphs for testing community detection algorithms. Phys. Rev. E **78**(4), 046110 (2008)

Part III
Electric Vehicle: A Game-Changing Technology for Future of Interdependent Networks

Chapter 8
Barriers Towards Widespread Adoption of V2G Technology in Smart Grid Environment: From Laboratories to Commercialization

Nadia Adnan, Shahrina Md Nordin and Othman Mohammed Althawadi

8.1 Introduction

The secretion of greenhouse gasses (GHGs) is the redundant by-product generally connected with the burning of fossil fuel for energy requirements [1, 2]. The major thrust of the world is going through crucial issues like energy scarcity, air pollution and the secretion of greenhouse gas (GHG) [3]. Electric Vehicles, which utilize both the electrical and internal combustion engines for propulsion purposes, materialize to show potential prospect [3, 4]. Furthermore, PHEVs are widely predictable as an answer that will decrease the damaging effect on the climate and lessen the carbon secretion [5]. Therefore, this type of vehicle offers a lead in the search to reduce carbon secretion by as much as 30–50% and be able to attain 40–60% upgrading in fuel efficiency [1]. While C. Barbarossa et al. [6] stated that it is the real fact they are going to be impartially on the lower side. A number of researchers have proven that a large amount of decrease in greenhouse gasses secretion and in the accumulating dependence on oil could be proficient by the electrification of the transportation sectors, which further needs suitable understanding and acceptance from the consumer's point of view [7–9]. Electric vehicle (EV) has experienced major transformations in the last few decades. The achievement of the smart electric grid with the addition of renewable energy

N. Adnan (✉) · S. Md Nordin
Department of Management and Humanities, Universiti Teknologi PETRONAS (UTP),
Bandar Seri Iskandar, 32610 Perak, Malaysia
e-mail: nadia.adnan233@gmail.com

S. Md Nordin
e-mail: shahrina_mnordin@petronas.com.my

O. M. Althawadi
College of Business and Economics, Qatar University (QU), Building Al Jamia Street Al
Dafna Area, Doha 2713, Qatar
e-mail: oalthawadi@qu.edu.qa

© Springer International Publishing AG, part of Springer Nature 2018
M. H. Amini et al. (eds.), *Sustainable Interdependent Networks*,
Studies in Systems, Decision and Control 145,
https://doi.org/10.1007/978-3-319-74412-4_8

exclusively depends on the extensive diffusion of EV for a carbon-free and sustainable transportation zone. Various efforts are being understood to decrease the secretion in the transportation zone. The main centre of attention is to sequence towards the implication of clean technology, which aims to reduce the GHG secretion and increase the vehicle performance [10]. The electrifying transportation is one of the promising approaches with many profits. By using this approach, EVs could get better energy security by diversifying energy sources hence foster economic growth. The EVs was developed to overcome the disadvantages of internal combustion engine ICE vehicles [11]. Lawful agencies around the entire earth are implementing different initiatives, strategies and agendas for broader EVs uptake [12]. The determinations appeared to pay off as EVs flinch to advance the acceptance among the public.

Stimulating transportation sectors show a potential method to progress the environmental-related changes and problems associated with that transportation sector [13]. The acceptance of electric vehicle into the market has introduced major impacts on a variety of fields, more than ever the power grid [14]. A number of policies have been implemented towards the promotion of electric vehicle operation with the rising inclination of electric vehicle adoption in the present eras has been fulfilling the need of thoughtful consumers [15]. The frequent development of electric vehicle power grid, battery and charger technologies has added development in the direction of electric vehicle technologies for broader acceptance among consumers [7]. Regardless of the ecological and economic compensation, EVs charging commences adverse influences on the accessible power grid procedure [14]. Due to lack of suitable charging management, plans can be tailored in order to overcome the issue related to EVs [16]. Furthermore, PHEVs adding in the smart grid can bring a number of conceivable probabilities particularly from the viewpoint of the innovation based on vehicle-to-grid (V2G) which is a thoughtful way for the renewable energy intermittency issue [17]. This research study is to look the latest expansion on EVs technologies, and their impacts and opportunities delivered by electric vehicle innovation [7]. According to Adnan et al. [1], EVs went through a series of technological developments before achieved the recent popularity. The repeated growth of EV technologies is important to arrange and struggle with the leading EV wider consumption and observation to be found on improving technologies, particularly oil price is kept rising, more automakers were dedicated to vehicle electrification [5]. Therefore, these components experience major shift along the EV improvement process. EVs can be hierarchical in a small number of types based on the vehicle hybridization ratio, which are hybrid electric vehicles (HEVs), plug-in hybrid electric vehicles (PHEVs) and battery electric vehicles (BEVs). Whereas acceptance of EVs gives green sustainability, PHEVs are claimed to be green and ecological-pleasant since PHEVs have fewer tailpipe emissions [7]. However, PHEVs usage is based on electricity produced through power grid in order to charge their batteries and the power generation procedure does not generate GHG emissions [4]. Currently, the transportation sectors heavily consume gasoline or petrol for their impulsion [9]. Nearby there is no interconnection among the transportation sector and power grid throughout that period. Conversely, the condition alterations with the wide adoption of PHEV into the marketplace since PHEV can be plugged into the power grid to take a distribution of energy [7]. When the interrelation among power grid and transport sector occurs, extensive investigations have been approved out to study the harmful impacts of PHEV charging on the power grid, which is previously addressed in [18]. Other than challenges, PHEV consumption, in reality, bring many opportunities to the parallels residential smart grid [19, 20]. Some interesting opportunities brought by the PHEV deployment in the smart grid are the V2G skill and addition of renewable energy sources and PHEV [21].

This research is to ensure that the return to higher oil prices in 2008 with the resource hovering at more than $140 per barrel for a short time has convinced some consumers to switch permanently from gas-guzzling behemoths to more energy-efficient vehicles. However, it also appears that the motivations for this switch were not detailed economic analyses, but simple reactions to sharp increases in the price of fuel. Most people apparently remain unable or perhaps unwilling to conduct careful economic calculations of the cars they buy a trend that will take more than higher oil prices to change.

8.2 Contribution

The history of other energy transitions implies that these "socio-technical" obstacles may be just as important to any V2G transition and, perhaps because they are often harder to identify, more difficult to overcome. Because no commercially feasible PHEVs implication presently available on the market, our assessment has the benefit of informing policy makers before they commit to a predetermined technological pathway. Given that energy technologies such as refineries and power stations require extremely large capital expenditures, the infrastructure built today will remain in operation for 30–40 years. By identifying a range of barriers to PHEVs and an eventual V2G transition now, we can help inform policymakers early in the process and perhaps avoid spending huge amounts of money on a promising technological pathway that fails to deliver results.

8.3 Literature Review

What is V2G? Vehicle-to-grid (V2G) describe as the bi-directional communication set-up, the application of V2G perception will develop reasonable by regulatory and dealing with energy conversation among the PHEV battery and power grid [17]. PHEV obtains indicating from the power grid when the charging level of PHEV battery is comparatively lessened [9]. In the line of V2G process, the charging level of the PHEV battery is uninterruptedly observed and allowable to be liquidated towards the smart grid [14]. Whereas PHEVs can be measured as energetic dispersed energy loadings [5]. Consequently, suitable V2G controlling for the large fleet of PHEV is significant to attain numerous assistances, such as dynamic power parameter, responsive power parameter and subsidiary service sustenance [22]. The V2G innovation received more attention than conventional due to the expansion of the bi-directional PHEV charges, which have permitted with the two-ways communication and energy conversation among power grid and PHEVs [6]. The innovation of V2G might bring numerous benefits towards the power grid with the proper V2G regulator and control [19]. However, the large fleet of grid-connected PHEVs will lead in the direction of different consequence in many undefined restraints towards the power grid, such as the dissimilar state of charging levels of PHEV batteries and the energetic prospect of PHEV assembly [7]. In command to accomplish the large fleet of PHEVs, component pledge optimization system is cast-off for development and regulatory the energy flow among PHEVs and power grid [23].

Problem identification for this research study is an important part in order to assess existing information and identify the problem that the previous researchers trying to address. The objective is to classify the problems of V2G adoption and their root cause, measure their influences and ensure the biggest problems which need to addressed as a priority. In the line of this research study, the technology road plotting of a smart battery charging infrastructure for PHEVs in order to aim they integrate this within the smart grids. The battery charging procedure is measured by a suitable regulator procedure, pointing to reserve the battery-operated life cycle [23, 24]. The foremost topographies of the gear are the modification of the power excellence deprivation and the bi-directional process, as a system of the grid to vehicles G2V and as a vehicle-to-grid (V2G). Therefore, the V2G mode of operation will be one of the foremost topographies of the smart grids, together with the collaboration with the electronic power grid to intensification constancy and to purpose as a dispersed energy storage system (ESS) [9]. Consequently, PHEVs could show an essential part in decarbonizing street transport in the near upcoming. To found the suitable policies for research and development (R&D) is obligatory. Rendering to Hoang et al. [17], technology road mapping is an appropriate means to shape up planned and long-term strategies by measuring possibly disrupting innovation skills and market variations. Thus, the unbiased approach of this research study makes consumers considerate for smart grid technology. In particular, this research focuses on the application of smart grids in the PHEV and V2G system adoption.

8.3.1 Theorizing V2G and PHEVs with the Practical Experiments Ahead

Furthermost, contemporary vehicles pay more attention towards ICE internal combustion engines because its start rapidly and deliver power as early as the as

Table 8.1 Difference between conventional and non-conventional vehicles

Difference between conventional and non-conventional vehicles		
Type of vehicles	Machine engine	Benefits
V2G-enabled (PHEV)	It carries a large battery and electric motor with small/ultimately no (ICE) internal combustion engine with additional V2G capabilities	It carries all features and benefit of PHEV with additional features of power sending facilities to the grid
Plug-in hybrid vehicle (PHEV)	It carries large battery and electric motor with smaller ICE internal combustion engine	It can recharge at night in order to apprehension HEV assistances added an AER all-electric range variable from 20 to 60 miles (about 30–100 km)
Hybrid electric vehicle (HEV)	It carries the entire features ICE internal combustion engine with separate electric motors	It also carries reformative braking and fuel reserves
Conventional vehicles (ICE vehicle)	It brings the features of ICE internal combustion engine	The main features of carries by ICE it carries rapid starting, comparatively fast speeding up and powerful

vehicles drivers need it. However, they function unproductively and discarded energy when lazing [6]. Dissimilarly plug-in hybrid electric vehicles (PHEVs), which have seen profitable accomplishment such as the Toyota Prius, Honda Insight, the Honda Civic Hybrid and many others, these varieties of product in the vehicle have additionally added a high functional battery and electric motor to vehicles, which contain ICE as well [7]. In the line of progressive power electronics and computer controls with conventional and electric drive vehicles, PHEVs function more proficiently than those that run on ICE unaided and lessen releases CO_2 [18]. They lessen fuel usage because they employ the electric motor frequently (particularly in the slow circulation of traffic), as they shut down the ICE once the vehicle has motionless for a prearranged sum of time, and since they evoke else they can cast-off dynamic energy throughout decelerating [1, 5, 7, 18]. Whereas Table 8.1 illustrates the difference between conventional and non-conventional vehicles.

PHEVs transmit topographies, which comprise all the feature of PHEV innovation additionally it carries features, which contain a larger battery. Furthermost, Rahman et al. [9] PHEV prototypes comprise a larger battery which is proficient of driving the vehicle about 20 and 60 miles (nearly 30–100 km) consuming electric power alone. In the line of this research Kempton and Tomić [25], stated the example of taking an initiative about advertising the Chevrolet Volt, an all-electric vehicle (AEV) that can function up to 40 miles on the household present without revitalizing [7].

In conclusion, a vehicle capable of "vehicle-to-grid" (V2G) interface, occasionally mentioned to as "mobile energy" or "smart charging", companions a vehicle with the current electric utility system [6, 23, 25]. Automobiles must retain three fundamentals function in the V2G establishment: a power assemblage power grid, a control and/or communication device that consents the grid operatives admittance to the battery-operated and accuracy on board the vehicle to path energy flows [14]. This intellectual, two-way interaction between the power grid and the car allows utilities to achieve power possessions in a better way, and it approves car owners to obtain cash by selling influence back to the grid [6].

The V2G and PHEVs schemes are consequently familiarly consistent. The PHEVs have the chance to develop not only vehicles but also mobile, self-sufficient incomes that can accomplish electricity flow and relocate the essential need for electric efficacy substructure [22]. The V2G cars can decrease the era price of the PHEVs, thus creation them extra striking, and if V2G upsurges the market perspective of PHEVs, the advantages of PHEV usage will also upsurge [26]. In the line of the context, the benefits and obstacles opposite the PHEVs continue unified with those who are the opposite V2G, which clarifies our argument, which is based on the adoption of V2G [6, 14].

8.3.2 The Benefit of V2G Transition

The concept of V2G stimulates promoters since it offers joint advantages to the transport sectors and the PHEV as well as the electronic schemes [9]. V2G has served the benefit that is associated with the reduction of gasoline usage, consolidation of the budget; enhance the national security, reducing damaging on gasoline substructure and refining the natural atmosphere [23]. Furthermore, it could add capacity to the power-driven grid through peak times deprived of the necessity for the efficacy business to shape new influence towards the plants [6]. The instant effect of extensive use of PHEVs might be lesser petroleum price. Upsurges in petroleum prices in 2008 and 2009 arose not only since the crude oil purchase price climbs, however, due to the purifying ability and deficiencies [19]. The world economist has a theoretical level that the crude oil could be free, but high prices for fuel would still exist because plants cannot deliver sufficient petrol [21]. In this research, superior market diffusion of PHEVs might directly restrain petroleum usage, facilitation plant scarcities and likely reduce prices. Conversely, in the long term, V2G will reduce the oil import through a large number of penetrations of V2G towards the transportation sector, which gives them additional benefits [6, 17].

However, the evolution towards the adoption of PHEV/V2G idea might also deal main environmental advantages [27, 28]. In this study, it is stated that the Minnesota Pollution Control Agency [29] intended that per mile releases CO_2 that is expectable the PHEVs might drop about 60–70% when associated with the conventional cars. In the line of this study documented that PHEVs decrease carbon dioxide's (CO_2) by 59–66%, nitrogen oxides (N_2O) from 48 to 80% and particular matter from 66 to 76%. The supposition that the PHEVs had an AEV range between 20 and 60 miles was phased in for light-duty cars were mechanical with power from a range of 60% coal and 40% gust [27]. The environmental impact of V2G vehicles contrariwise, but it found that even when powered totally by coal-fired electricity, PHEVs still produce around 25% GHG emissions per mile than do conventional vehicles [27, 30, 31]. This study underlined that the valuation importantly underestimates the GHG reduce the potential for PHEVs [13]. The emission might be lesser since the usefulness of collections might comprise some lesser carbon producers, which is based on renewables and cogeneration components and would not consist of 100% coal-fired generators, as the study assumed [32]. Furthermore, authors discuss the consumer perspective towards the adoption of PHEVs and V2G systems.

8.3.3 Consumer Attitudes and Motivations

Consumer attitude and inclinations for PHEVs must be measured in emerging marketplace portion [33]. PHEVs are essential not only dazed the technical glitches opposite the battery technology but also societal concerns connected to consumers in a directive to attain profitable achievement [18]. However, the consumer adoption is critical to the enduring accomplishment towards the sustainable transportation sectors [7]. Though consumers incline to be resistant in order to the adoption of PHEVs that is measured as unacquainted or unverified [5]. Thus, failure by PHEV producers and policy makers to classify and dazed consumer problems which may result in continual low acceptance of PHEVs long after the technological problems are determined. In order to make the consumer accept the adoption of PHEV via a V2G system [17]. The researcher has used as the theory of planned behaviour (TPB) by Ajzen [34] which clarifies the factors influencing the consumer behaviour [6]. Based on the TPB, the main factors that predict the consumer behavioural intention are attitudes, it explains information and knowledge, whereas subjective norms explain that the consumer to be in certain acceptable level, which is perceived societal, pressure which indicates to perform and not to perform a certain behaviour which impacts the adoption level [9]. In the line of this book chapter, consumer adoption is considered an intention to adopt, use, PHEVs [34]. The basic cognitive of the TPB is that movements which are selected founded on an investigation of the substitutes over the optimum consequence are attained [7].

Researchers investigate that some mutual obstacles towards the adoption of any new knowledge that comprise the deficiency of knowledge by educating the possible adopters, which is based on high preliminary costs and low-risk of acceptance [1, 9]. A research stated by Adnan et al. [1] designates that consumer acceptance of PHEVs is inadequate partially due to apparent hazards with new products and interchanges between vehicle fuel competence, price and size. The consumer's awareness of hazard is founded by a knowledge, reactions are based on the non-technical aspect towards the adoption of PHEV which is done with the help of V2G [19]. Generally, the social media or the mass media and social networks frequently affect standards which are based on consumer choice towards the adoption of PHEVs specifically based on V2G technology [9, 19]. In relationships with economic assistances, consumers are added which probably to indicate decisions that make the most of the utility built on their inclinations, information of substitutes and low-priced [19]. The preliminary cost of PHEVs is pointedly higher when it is associated with a petrol power-driven ICE vehicle and this cost growth is higher due to the size of the battery as well as the range of the vehicles [7]. Furthermore, the non-financial explanations, particularly those related to environment and green energy can affect consumers' decisions to buying a PHEV [8]. In future, the potential for PHEVs to generate social welfares by dropping petroleum ingesting and GHG emissions can application to particular consumers. Ecological ethics are influential forecasters of convinced consumer schedules and absolutely affect willingness to involve in activities that defend the environment [7, 31, 35]. Heffner et al. [36] originate that, to this collection of consumers, who illustration high levels of environmental consciousness, indicating a PHEV signifies thoughts associated to one's specific which used to connect benefits and standards. Reviewing the PHEV purchases in Falvo et al. [22] originate that ecologists are extra probable to acquisition the PHEVs associated to non-environmentalists. Similarly, Al-Alawi et al. [4] originate that communal predilections for environmental fineness and energy security were a major factor for consumer adoption of PHEVs. Rahman et al. [9] determined that social favourites amplified PHEV sales more than rising gas prices or tax incentives. Besides, the traditional tendencies in technology adoption suggest that though innovation is essentially attractive to insufficient initial adopters, counting seeds and technology supporters, the widespread of consumers will endure close-minded about the innovation [37]. However, this small segment of primary adopters has significantly attitudes to innovation and is likely to adopt new innovation [36]. In the line of this research, some consumers are uncomfortable with technological change and uncertainty, and therefore are hesitant to accept innovations [38]. Heffner et al. [36] found that the people who show high levels of environmental awareness, choosing an EV symbolizes ideas related to one's individuality and is used to communicate interests and values. The majority of consumers, although making choices, stick to "ideas of traditional knowledge" somewhat than acceptance a new innovation [7]. In the line of the new eras, though there are cumulative motives to adopt PHEVs counting increasing and unstable petrol values, greenhouse gas emissions, increased need for imported petroleum, and the very high fuel budget of PHEV [8].

8.3.4 Barrier of V2G Transition Based on Social and Cultural Aspect

These potential benefits obviously supporters of the V2G idea, urging them to last effort on what they see as the main difficulties, specifical difficulties with battery

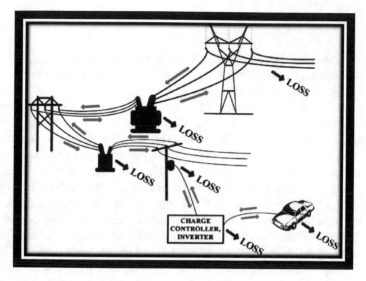

Fig. 8.1 Losses analysis of V2G

innovation and high prices associated with conventional IC cars [39]. However, power-driven problems to a PHEV transition noticeably happen, the interval of this object claims that researchers and lawful agency vital to classify other impairments [40]. Furthermost prominently, they necessity reflect that influence of a congregation of socio-technical deliberations, such as those that might ascend between consumers, those linking to commercial practices and checking the aspect of lawful agencies, and that business with communal struggle [8]. It might opportunity out that, smooth with practical glitches determined the V2G idea may not improvement extensive adoption [17]. To assistance comprehend the glitches; we associate V2G apprehensions to those knowledgeable in the overview of other non-traditional energy innovation [27]. Furthermore, it is more understandable that energy transformation losses due to efficiency-related issues while applying V2G in order to tackle the power source. However, this is losses occurs each time when the energy stored or transmitted. Whereas in PHEV, the quantity of losses differs and can be very large which losses in the internal combustion engine, in the line with smaller losses which can be in power electronic devices and electric energies. In this study, PHEV-grid or EV-grid are inter-connected, it is understandable that, once energy goes over the numerous phases of storing, alteration, parameter and diffusion, each stage underwrites to losses. Additionally, V2G procedure rights that energy stream can be overturned, and kept energy can be sent back towards the grid [41]. Lastly, Fig. 8.1 shows the V2G losses analysis, which could be other barriers towards the adoption of the V2G system. The storage is not at the higher level.

8.3.5 Consumers Concerns

There are numerous issues related to customers, which could be a lack of awareness among consumers [10] towards the benefit of PHEVs, particularly in the arena of effective management of energy usage and the environmental benefit. The main crucial point for the customers is that the vehicles depend upon the recharging services, during long journeys the risk of being out of energy while driving the vehicles. The recharging necessities need to be site autonomous, which allows the drivers to charge the vehicles regardless of the particular place. Automatic identification of cars essential that need to be possible [8]. Inadequate charging infrastructure will prime range of apprehension. Furthermore, the problem leads towards the stealing and damage of consumer charging infrastructure. Consumer's anxiety on unified payment unavailability of fund and security of payments and money transactions. Numerous charging service operators lead to interoperability problems [21]. The consumer of PHEVs should be mindful of the responsibility position of the PHEV sequences. Additionally, it will lead the inadequate size of the parking portion which central to jam and consumption of time for the car's execution of the grid transactions. Inappropriate forecasting concern ascends that create its challenges in order to regulate the suitable time for an assumed car to purchase or sell power control [28]. Intellectual preparation of PHEV's is crucial to empower the consumers to contribute towards the management demand side PHEVs vehicles operators towards the need evidence about the accessibility of electrical energy system and the costs for the dissimilar substitutes that they can select from at any assumed period [37].

8.4 Conclusions and Future Direction for Transportation Policymakers

An evolution of V2G innovation has ample to the proposition. The phenomena of reducing gasoline usage would aid defending the oil introducing markets from gasoline cost points and shudders to the worldwide marketplace, attractive nationwide safety and justifying the transmission of capital to oil-producing nations [42]. It might also importantly recover the superiority of the atmosphere, transferring injurious CO_2 emissions and the healthiness, environmental and climate-changed compensations they transport with them. Furthermore, PHEVs is the essential predecessor to V2G innovation, proposal car driver possible price investments from their use of power as a fuel in its place of petrol, and they might importantly progress the economic routine of power-driven efficacy businesses, principally those that usage renewable energy producers such as wind turbines and solar panels.

Though the supports of such a changeover have been extensively documented, which has not up until now been attained, possibly since the impairments opposite such innovation continue concurrently procedural and communal, particularly for the PHEV, firstly the link in a V2G evolution. Impairments linking to purchaser adoption, the historic abhorrence towards innovation, and enthusiastic confrontation from investors in the prevailing infrastructure may be noteworthy. V2G innovations and PHEVs may understand refusal from consumers since of their high original price, a thoughtful obstacle since that most persons do not reduction the investments from energy-efficient innovations as do monetary specialists. Drivers will probably be uninformed of how their driving designs and behaviours insignificantly affect V2G PHEV enactment, showing irritation and hindrance if their new vehicles do not achieve exactly as predicted. Additional thoughtful confrontation may come after vehicle producers, oil corporations and restoration industries that have defeated billions of dollars into the source and manufacture substructure for predictable cars. One would imagine these influential businesses to utilize huge inspiration with legislators and the community to uphold the position.

This particular research study absorbed on the observations and attitudes of a technically minded collection towards PHEVs. The forthcoming research study will associate the perception and attitude with the consumer in order to deliver understanding on in what way dissimilar categories of customers recognize PHEVs and V2G system in the direction to delivers the high point consumer's correspondences and alterations among the two dissimilar customer clutches [9, 27]. However, the ownership cost of V2G and PHEVs deliberated in this research articles, which also leads towards the future prospects of this work. In another part of the worlds as Europe, the gasoline price is higher compared to another part of the developed nation [2, 28, 43]. Consequently, deprived of other inducements, customers will probably to be additionally interested in obtaining PHEVs in Europe than in the USA [27]. However, in the developing nations, customers are more willing to buy V2G-enabled PHEVs system when they aware about the harmful effect of CO_2 emission which is an injurious outcome to the environment [42]. The customer who takes adoption of PHEVs and V2G system can enhance their lifestyle. Whereas Fig. 8.2 reveals the virtual power plant structure in order to store the grid. Virtual power plant not only compacts with the source side, nonetheless, it also benefits to accomplish the plea and safeguard the grid dependability through plea reaction in the real period [41].

A number of institutional barriers also need be addressed [13] such as: (1) necessity of adequate standards for upcoming penetration of electric vehicles with V2G services, (2) lack of mass production of vehicles with V2G capability, (3) unavailability of regulation services rates at the retail level, (4) regulation signal is not broadcasted by all Independent System Operators (ISOs) and (5) absence of vehicles aggregators to control individual vehicles as well as multiple fleets.

Furthermore, if we receive that PHEVs and V2G innovation have noteworthy compensations, but continue hindered by socio-technical difficulties, then the research and development trails need some modification. Whereas the continued research effort will endorse the improved quality of battery and it is also related to the control innovations. Certainly, researcher deliberate that better-quality batteries, for instance, can assistance lessen some of the impairments researcher designates.

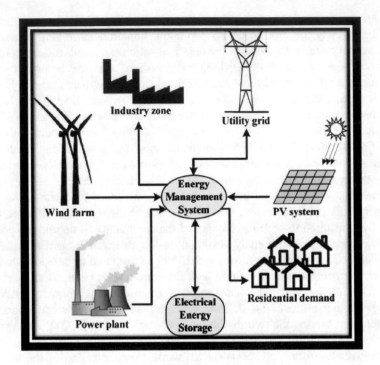

Fig. 8.2 Virtual power plant

Nevertheless, researcher purely reminder that effort to recover the mechanical enactment of hardware must be attached with efforts to overwhelmed financial, behavioural, ethnic and infrastructural difficulties. However, these categories of blockades do not appropriate effortlessly into the outdated research and development groups and continue intensely entrenched in the social and official material. Overpowering them may necessitate a considerable exertion that presently escapes many conversations.

8.5 Acknowledgement

The authors would like to acknowledge Universiti Teknologi PETRONAS (UTP), Department of Management and Humanities and Graduate Assistantship (UTP-PG/ADM_PGC November-2014 AL1406182) scheme to facilitate this research study. The authors would also like to thank the reviewers for their valuable suggestion in order enhances the manuscript.

References

1. N. Adnan, S.M. Nordin, I. Rahman, P. Vasant, M.A. Noor, An overview of electric vehicle technology: a vision towards sustainable. Green Mark. Environ. Responsib. Mod. Corp. 216 (2017)
2. M.H. Amini, M.P. Moghaddam, O. Karabasoglu, Simultaneous allocation of electric vehicles' parking lots and distributed renewable resources in smart power distribution networks. Sustain. Cities Soc. **28**, 332–342 (2017)
3. J. Axsen, K.S. Kurani, Hybrid, plug-in hybrid, or electric—What do car buyers want? Energy Policy **61**, 532–543 (2013)
4. B.M. Al-Alawi, T.H. Bradley, Review of hybrid, plug-in hybrid, and electric vehicle market modeling studies. Renew. Sustain. Energy Rev. **21**, 190–203 (2013)
5. N. Adnan, S.M. Nordin, I. Rahman, P.M. Vasant, A. Noor, A comprehensive review on theoretical framework-based electric vehicle consumer adoption research. Int. J. Energy Res. (2016)
6. C. Barbarossa, S.C. Beckmann, P. De Pelsmacker, I. Moons, W. Gwozdz, A self-identity based model of electric car adoption intention: a cross-cultural comparative study. J. Environ. Psychol. **42**, 149–160 (2015)
7. N. Adnan, S.M. Nordin, I. Rahman, Adoption of PHEV/EV in Malaysia: a critical review on predicting consumer behaviour. Renew. Sustain. Energy Rev. **72**, 849–862 (2017)
8. N. Adnan, P.M. Vasant, I. Rahman, A. Noor, Adoption of plug-in hybrid electric vehicle among Malaysian consumers. Ind. Eng. Manage. **2016** (2016)
9. I. Rahman, P.M. Vasant, B.S.M. Singh, M. Abdullah-Al-Wadud, N. Adnan, Review of recent trends in optimization techniques for plug-in hybrid, and electric vehicle charging infrastructures. Renew. Sustain. Energy Rev **58**, 1039–1047 (2016)
10. N. Adnan, S.M. Nordin, I. Rahman, A. Noor, Adoption of green fertilizer technology among paddy farmers: a possible solution for Malaysian food security. Land Use Policy **63**, 38–52 (2017)
11. M. Coffman, P. Bernstein, S. Wee, Electric vehicles revisited: a review of factors that affect adoption. Transp. Rev. **37**, 79–93 (2017)
12. T.A. Becker, I. Sidhu, B. Tenderich, Electric vehicles in the United States: a new model with forecasts to 2030. Center Entrep. Technol. Univ. Calif. Berkeley **24** (2009)
13. R.J. Javid, A. Nejat, A comprehensive model of regional electric vehicle adoption and penetration. Transp. Policy **54**, 30–42 (2017)
14. W. Kempton, J. Tomić, Vehicle-to-grid power implementation: from stabilizing the grid to supporting large-scale renewable energy. J. Power Sources **144**, 280–294 6/1/ 2005
15. N. Adnan, S.M. Nordin, A.N. Noor, A comparative study on GFT adoption behaviour among Malaysian paddy farmers. Green Mark. Environ. Responsib. Mod. Corp. 239–263 (2017)
16. M. Kassie, M. Jaleta, B. Shiferaw, F. Mmbando, M. Mekuria, Adoption of interrelated sustainable agricultural practices in smallholder systems: evidence from rural Tanzania. Technol. Forecast. Soc. Chang. **80**, 525–540 (2013)
17. D.T. Hoang, P. Wang, D. Niyato, E. Hossain, Charging and discharging of plug-in electric vehicles (PEVs) in vehicle-to-grid (V2G) systems: a cyber insurance-based model (2017). arXiv:1701.00958
18. S.M.N.N. Adnan, I. Rahman, P. Vasant, M.A. Noor, An integrative approach to study on consumer behavior towards plug-in hybrid electric vehicles revolution: consumer behavior towards plug-in hybrid electric vehicles. Appl. Behav. Econ. Res. Trends, IGI Global 185 (2017)
19. M.L. Tuballa, M.L. Abundo, A review of the development of smart grid technologies. Renew. Sustain. Energy Rev. **59**, 710–725 (2016)
20. M. Amini, A. Islam, Allocation of electric vehicles' parking lots in distribution network, in *2014 IEEE PES Innovative Smart Grid Technologies Conference (ISGT)*, pp. 1–5 (2014)

21. P.D. Lund, Clean energy systems as mainstream energy options. Int. J. Energy Res. **40**, 4–12 (2016)
22. M.C. Falvo, L. Martirano, D. Sbordone, E. Bocci, Technologies for smart grids: a brief review, in *2013 12th International Conference on Environment and Electrical Engineering (EEEIC)*, pp. 369–375 (2013)
23. K.G. Boroojeni, M.H. Amini, S. Iyengar, *Smart Grids: Security and Privacy Issues* (Springer, 2016)
24. N. Adnan, S.M. Nordin, I. Rahman, A.M. Rasli, A new era of sustainable transport: an experimental examination on forecasting adoption behavior of EVs among Malaysian consumer. Transp. Res. Part A Policy Pract. **103**, 279–295 (2017)
25. W. Kempton, J. Tomić, Vehicle-to-grid power fundamentals: calculating capacity and net revenue. J. Power Sources **144**, 268–279 (2005)
26. O. Ellabban, H. Abu-Rub, F. Blaabjerg, Renewable energy resources: current status, future prospects and their enabling technology. Renew. Sustain. Energy Rev. **39**, 748–764 (2014)
27. S.S. Hosseini, A. Badri, M. Parvania, The plug-in electric vehicles for power system applications: the vehicle to grid (V2G) concept, in *2012 IEEE International Energy Conference and Exhibition (ENERGYCON)*, pp. 1101–1106 (2012)
28. H. Turton, F. Moura, Vehicle-to-grid systems for sustainable development: an integrated energy analysis. Technol. Forecast. Soc. Chang. **75**, 1091–1108 (2008)
29. J.L. Martinez, Environmental pollution by antibiotics and by antibiotic resistance determinants. Environ. Pollut. **157**(11), 2893–2902 (2009)
30. M.H. Khooban, T. Niknam, M. Sha-Sadeghi, Speed control of electrical vehicles: a time-varying proportional–integral controller-based type-2 fuzzy logic. IET Sci. Measur. Technol. (2016)
31. M. Amini, A.I. Sarwat, Optimal reliability-based placement of plug-in electric vehicles in smart distribution network. Int. J. Energy Sci. **4** (2014)
32. F. Kley, C. Lerch, D. Dallinger, New business models for electric cars—a holistic approach. Energy Policy **39**, 3392–3403 (2011)
33. N. Adnan, S.M. Nordin, A.N. Noor, A comparative study on GFT adoption behaviour among Malaysian paddy farmers, in *Green Marketing and Environmental Responsibility in Modern Corporations*, ed: IGI Global, pp. 239–263 (2017)
34. I. Ajzen, The theory of planned behaviour. Organ. Behav. Hum. Decis. Process. **50**(2), 179–211 (1991). https://doi.org/10.1016/0749-5978(91)90020-T
35. M. Amini, B. Nabi, M.P. Moghaddam, S. Mortazavi, Evaluating the effect of demand response programs and fuel cost on PHEV owners behavior, a mathematical approach, in *2012 2nd Iranian Conference on Smart Grids (ICSG)*, pp. 1–6 (2012)
36. R. Heffner, K.S. Kurani, T.S. Turrentine, *Symbolism in early markets for hybrid electric vehicles: Institute of Transportation Studies* (University of California, Davis, 2007)
37. I. Rahman, P. Vasant, B.S.M. Singh, M. Abdullah-Al-Wadud, Swarm intelligence-based optimization for PHEV charging stations. Handb. Res. Swarm Intell. Eng. **374** (2015)
38. R.R. Heffner, K.S. Kurani, T. Turrentine, Effects of vehicle image in gasoline-hybrid electric vehicles. Inst. Transp. Stud. (2005)
39. O. Egbue, S. Long, Barriers to widespread adoption of electric vehicles: an analysis of consumer attitudes and perceptions. Energy policy **48**, 717–729 (2012)
40. Z. Rezvani, J. Jansson, J. Bodin, Advances in consumer electric vehicle adoption research: a review and research agenda. Transp. Res. Part D Transp. Environ. **34**, 122–136 (2015)
41. E.S. Dehaghani, S.S. Williamson, On the inefficiency of vehicle-to-grid (V2G) power flow: potential barriers and possible research directions, pp. 1–5
42. D.B. Richardson, Electric vehicles and the electric grid: a review of modeling approaches, impacts, and renewable energy integration. Renew. Sustain. Energy Rev. **19**, 247–254 (2013)
43. N. Adnan, S.M. Nordin, I. Rahman, M.H. Amini, A market modeling review study on predicting Malaysian consumer behavior towards widespread adoption of PHEV/EV. Environ. Sci. Pollut. Res. 1–21 (2017)

Chapter 9
Plug-in Electric Vehicle Charging Optimization Using Bio-Inspired Computational Intelligence Methods

Imran Rahman and Junita Mohamad-Saleh

9.1 Introduction

Transports which gain most of their energy from the power grid (which includes all-electric vehicles and plug-in hybrid vehicles) have attained noteworthy market diffusion over the past few years [1, 2]. Such transports, commonly mentioned as plug-in electric vehicles (PEVs), lessen the fossil fuel consumption and hence decrease the emissions which includes greenhouse gases [3]. As the number of PEVs are increasing, power system operation will turn out to be more complex [4]. For instance, if a large number of electric vehicles start charging after most people complete their evening commute, a new demand peak could result conceivably demanding ample new power generation capacity as well as ramping capability [5].

The influence of PEVs on the power system has been studied in a few works [6, 7]. Scheduling PEVs charging/discharging profiles is one of the solutions to mitigate the impact of PEVs on the power grid. This can be performed by combining numerous sets of PEVs for charging or discharging with different durations and start times such that grid constraints are properly maintained. Nevertheless, the aggregation of PEVs varies from the aggregation of more conventional power sources [7]. In particular, the temporal availability of PEVs with their precise location information is a significant constraint to study while aggregating PEVs for probable grid congestion management and planning. Therefore, finding suitable charging and discharging times of PEVs that do not disrupt grid constraints while preserving tolerable degrees of customer satisfaction is a challenging optimization problem.

I. Rahman · J. Mohamad-Saleh (✉)
School of Electrical & Electronic Engineering, Engineering Campus,
Universiti Sains Malaysia (USM), 14300 Nibong Tebal, Pulau Pinang, Malaysia
e-mail: jms@usm.my

I. Rahman
e-mail: imran.iutoic@gmail.com

© Springer International Publishing AG, part of Springer Nature 2018
M. H. Amini et al. (eds.), *Sustainable Interdependent Networks*,
Studies in Systems, Decision and Control 145,
https://doi.org/10.1007/978-3-319-74412-4_9

The earlier generations of electric vehicles are projected to be linked to the power grid only for battery charging. Nevertheless, as the innovative technology develops, the idea of V2G (vehicle-to-grid) will become practically applicable. V2G refers to the technique of injecting power to the grid while being connected to it from vehicle on-board battery. By the help of this system, next-generation PEVs will act as both generators and electric loads, i.e., PEVs will function as energy storage apparatus [8].

During the past few years, several works have been comprehended on bio-inspired optimization techniques. In most of the circumstances, optimum solutions are found by the help of hybrid techniques, particularly on actual-world problems. Earlier, cooperation was mostly comprehended between numerous CI methods. However, currently, gradual cooperation structures between general CI methods with exact tactics are suggested. Hybrid methods typically produce satisfactory results as they are capable of exploiting simultaneous advantages of both kinds of single techniques [9].

It is noteworthy to mention that some of the CI techniques have had their origins in pure science and engineering fields. However, there is a good prospective to explore various hybrid CI methods for solving power system problems as well as their associated theories for future enhancement [4]. Optimization techniques are usually studied as resolving scheduling problems related to PEV, power grid and consumer constraints. Nevertheless, there are other optimization issues like integration of PEV, sizing and placement of charging stations.

The remaining segments are organized as per the following way: Sect. 9.2 discusses the charging of plug-in electric vehicle (PEV), Sect. 9.3 highlights the PEV charging optimization issues, Sect. 9.4 describes the bio-inspired computational intelligence (CI) techniques, Sect. 9.5 discusses the applications of bio-inspired CI for PEV charging optimization, and finally, Sect. 9.6 concludes the chapter with future research directions.

9.2 Charging of Plug-in Electric Vehicle (PEV)

The PEV charging can be categorized into three different ways depending on the charging voltage level and locations. Figure 9.1 provides the schematic diagrams of different PEV charging alternatives.

The classifications of PEV based on the charging level are briefed below.

9.2.1 Level 1 Charging

Household-type socket-outlet is used for AC Level 1 slow charging. The best public choice for PEV charging is Level 1 type because of the usage of conventional

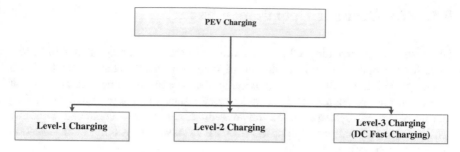

Fig. 9.1 Different types of PEV charging [10]

industrial/house socket. Level 1 type of charging is presently regarded as the key mode for small-sized PEVs like two wheelers [10].

9.2.2   Level 2 Charging

Slow charging from a household-type socket with an in-cable protection device is used for AC Level 2 charging which also permits the usage of conventional industrial/house socket [10]. Nonetheless, this type of charging delivers extra shield by adding an in-cable control box with a control pilot conductor amid the PEVs and control box or the plug.

9.2.3   Level 3 Charging (DC Fast Charging)

DC fast charging is used as an external charger. There are 2 sub-mode types of operation measured for the mode, specifically, the DC Level 1 (current <80 A, voltage <500 V, power = 40 kW) and the DC Level 2 (current <200 A, voltage <500 V, power = 100 kW) [10].

It is noteworthy to indicate that in Level 1 type, there is no physical communication between the charging point through the connector and PEVs. In Level 2 type, a pilot communication system can be added by permitting charging rate control. Meanwhile, Level 3 type is mostly utilized for fast charging purposes which is DC. A communication system is involved in Level 3 type of charging which allows the management of appropriate battery charging. Furthermore, in Level 2 and Level 3 types of charging, wireless communication networks can be utilized to interconnect with PEVs as well as to control the charging and discharging progression.

9.3   PEV Charging Optimization Issues

Scientists are now putting efforts to minimize certain parameters like life-cycle costs for the operation and installation of charging infrastructure, extra loads on charging station as well as to maximize the overall revenues, integration of renewable energy sources (RESs) and average state of charge (SoC) [11]. Furthermore, expert systems based on real-time simulation, enhanced charging schemes and optimal power allocation for PEVs have drawn ample consideration among the researchers.

Kulshrestha et al. in [12] proposed concurrent smart energy management of PHEVs for optimum usage of available power, charging duration and grid stability. Furthermore, consumer approval, loads conditions, SoC and storage capability are included. The benefits of using electric vehicles as energy storage for demand-side management are addressed by Pang et al. [13]. Herrera et al. developed hardware-in-loop simulation platform [14] for continuous (power systems and power electronics) as well as discrete (communication systems) constraints. Sizing optimization of the local energy storage (LES) for PEV charging was created within an overall cost-minimization agenda by the authors in [15]. The control mechanism of PHEV charging stations with LES facility was established. The results showed the superiority of proposed systems with optimized parameters during both the islanding and grid-connected modes as well as the transitional period with minimized voltage. Lu et al. [16] studied large-scale behavior of vehicle charging, deployment of charging infrastructure and driving pattern for PHEVs. Extensive analysis was performed considering PHEV charging and driving dataset and responded specific research questions on PHEV's interaction to traditional power grid network. Lastly, the researchers recommended the need of real-time driving data with global optimization techniques.

Tulpule et al. [17] formulated improved equivalent consumption minimization strategy (ECMS) for electric vehicle control by considering total energy consumption factor together with the constant SoC maintenance of vehicle battery. As a result, the SoC is dictated by ECMS at a persistent position with low consumption of fuel. Tehrani et al. [18] characterized the fast charging infrastructures' operation equipped with energy storage and RESs in order to optimize the charging pattern and retailing power to the existing grid by following the price variations to maximize the fitness function for the benefit of contributing to the electricity market. In [19], the fitness function was to reduce the total cost. In [20], the fitness function was to maximize the utilization of renewable energy and to minimize the charging cost. The optimization constraints are the size and charging rate of the battery. In [21], the objective was to guarantee fast charging duration of battery without the overheating. Moreover, in [22], the objective was profit maximization of PEVs. The authors provided different optimization objectives and certain system constraints with the simulated data. Conferring to the type of the optimization problems addressed, the authors suggested the estimation of distribution algorithm (EDA) to appropriately control numerous batteries charging/discharging from a bunch of electric vehicles.

The capability of PEVs to facilitate the renewable energy integration to the existing power system networks is possibly the most significant influence on power grid [23]. Placement of all-encompassing photovoltaic (PV) charging arrangement in an EV parking area was described by Neumann et al. [24]. Charging in PV parking lot and diverse business models considering solar energy were discussed by Rizzo et al. [25]. Environmental and socioeconomic influences of PV-based office charging infrastructure were addressed in [26]. The study specified the technical feasibility of establishing a PV-integrated office parking facility considering profits for the car owner as compared to the household facilities of charging. Authors stated that the consumer will receive the return of establishment and maintenance cost as well as profit within the lifespan of the PV panels. Birnie [27] introduced a solar collector integrated parking shade by encouraging the widespread installation of solar system module and concluded that the system will allow much more rapid payback period. In [28], Zhang et al. explained optimal control approaches to integrate both the PEVs and PV considering the existing power grid.

9.4   Bio-Inspired Computational Intelligence (CI)

Nature is certainly an enormous and potential source of motivation for solving complex problems in the domain of computer science as it shows very dynamic, diverse, complex and robust phenomenon [29]. It regularly finds the optimal solution to solve its problem keeping balance between exploration and exploitation. This is the thrust behind bio-inspired computational intelligence (CI). Bio-inspired CI techniques have the advantages as follows:

i. The stochastic and population-based nature of bio-inspired CI can significantly increase its search space and hence lessen the chance of trapping into local optima compared to the classical local optimization methods such as hill-climbing and gradient-based techniques which are employed on deterministic rules.

ii. Bio-inspired CI is focused toward better regions of a search space compared to the ineffective random search technique because it accurately uses fitness functions rather than function derivatives.

Bio-inspired CI techniques are very diverse and can be put into 3 categories: evolution based, swarm based and ecology based. It is important to mention that such categorization is not rigorous. Nonetheless, it is primarily done for the ease of discussions in the chapter. A brief taxonomy of bio-inspired CI techniques is shown in Fig. 9.2.

Črepinšek et al. [30] offered some basic guidelines to conduct any replications and comparisons of evolutionary computation-based algorithms for optimization.

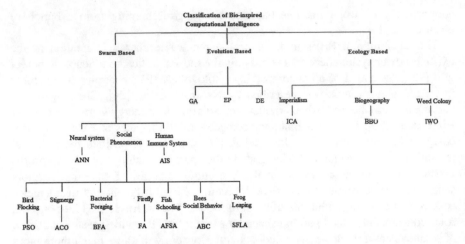

Fig. 9.2 Taxonomy of bio-inspired CI techniques

Moreover, the comparisons conducted should be based on suitable performance measures and able to show statistical significance of one approach over others. If the simulation study is not carried out with adequate caution, any statistical methods and performance measures cannot get rid of the problems adapted by inaccurate simulation replication. Defining suitable performance measures are the basis for algorithm comparisons. Hence, performance measures must be carefully defined and described. Exact replication cannot always be attained. All deviations must be stated. Any changes to the original experiment should be openly discussed along with a description of the inspiration for the changes, as well as any threats to the validities of the conclusions [31].

9.5 Applications of Bio-Inspired CI for PEV Charging Optimization

In this section, the bio-inspired CI techniques often used in the charging optimization are presented.

9.5.1 Charge Scheduling Optimization

By the appropriate establishment of intelligent scheduling techniques, smooth integration of PEVs onto the power grid can be attained. Smart scheduling techniques will assist to avoid cycling of bulky combustion plants, using costly fossil fuel peaking plant by introducing PEVs in power system networks.

Bio-inspired CI techniques are alternative optimization tools to deal with the non-smooth and complex power grid scheduling particularly with the latest deployment of electric vehicles with dubiety and well-regulated charging loads [32]. Moreover, the bio-inspired CI techniques uphold stochastic types of solutions as well as guide them toward optimal solutions through heuristic approaches. These methods are typically not certain toward global optimum achievement, but are generally resistant to high-dimensional, non-convex and nonlinear systems because of this process. Therefore, the mentioned techniques are widespread selections for elucidating the constraints and fitness functions which are not endlessly differentiable like binary charging/discharging scenarios considering an enormous PEV.

Authors used ant colony optimization (ACO) for transformer side charging scheduling of PEVs [33]. They concluded that the computation burden of ACO is relatively low and thus suitable for large-scale application. The simulation results relate and compare the load charging curve of PEV with the effect of load fluctuation. The authors also proposed an intelligent charging algorithm for electric vehicle charging services in reaction to TOU price. The aim was to improve the stress in power grid under peak demand and to meet the demand response requirements in regulated market. Authors in another work introduced a centralized scheduling policy for PEV charging using genetic algorithm (GA) to facilitate the size and complexity of the optimization [34]. The load curve shape remained relatively consistent. Thus, the algorithm attained statistically similar results from run to run. This validated the formulated optimization approach and algorithm for PEV charge scheduling.

In Soares et al., 3 variants of PSO techniques were formulated for comparison [35]. The authors concluded that, with the rise of decision variables, the overall computational complexity prolonged exponentially. From the results analysis, authors concluded that EPSO obtained better solution quality with reasonable execution time for the problem context (day ahead). Moreover, ant colony optimization (ACO) was utilized to improve the binary PSO for the optimization of unit commitment (UC)-related problem due the PEV load suggested by the authors in [36]. The scheduling was handled by binary PSO based on logical operators, and economic dispatch was solved through an improved ACO. The best cost per produced unit (CPU) procedure was implemented in the suggested technique to decrease the maximum required iteration in BPSO.

Roy and Govardhan examined the economic cost of UC with the generation of wind energy, emergency demand response, PEV charging and discharging using teaching-learning-based optimization (TLBO) algorithm [37]. The study indicated that the operational cost reduction is possible by the expansion of PEV battery capacity. Nonetheless, additional cost on transmission congesting and battery depletion exists among the challenges of PEV operation in power grid networks.

Proper charging scheduling is needed for upcoming PEV penetration in the vehicular network. There exists 'range anxiety' among the owners of PEVs. They are worried about the electric vehicle mileage because the on-board storage needs to be charged when the state of charge reaches a certain limit [38]. Uncoordinated fashioned PEV charging is the source of disturbances to the power grid, i.e., lines and transformers overload and voltage drops [39].

Hybrid optimization techniques perform a noticeable role in enhancing any search techniques. Hybrid techniques are used to combine the benefits of two or more algorithms, while concurrently trying to curtail any considerable drawbacks [40]. Overall, the effect of hybrid techniques can typically make some developments considering solution accuracy and computational complexity [41].

9.5.2 Optimal Charging Strategy

One of the most recent charging strategies of PEVs is smart charging [42]. The awareness behind this smart charging is based on PEV charging during the most advantageous scenarios when the electricity demand and price is the lowest with surplus capacity [43].

Authors in [44] used artificial immune system (AIS) and tangent vector (TV) technique for PEV recharging policy of IEEE 34-bus distribution system. The results of the TV-based optimization method demonstrated loss reduction with a lesser computational complexity by the help of random search process. Nevertheless, the authors suggested simulation for larger distribution systems as a future research.

Authors in [38] did a trade-off between power management strategy of stochastic optimal PEV and electrochemistry-based model of anode-side resistive film formation in lithium-ion batteries using a non-dominated sorting genetic algorithm (NSGA) in the formation of a Pareto front. Authors applied NSGA-II to the plug-in EV model in order to find its optimum charging patterns. After comparing various solutions from the Pareto front, the authors suggested that consumer should preferably charge a plug-in EV rapidly during off-peak hours and just before the onset of traveling to efficiently reduce energy costs and battery degradation.

Different PSO variants were utilized to optimize other PEV charging-related parameters. In [39], the fitness function was to maximize the average SoC in terms of the battery capacity, energy cost and remaining PEV charging time which is very nonlinear in nature and tough to resolve by traditional optimization techniques. The authors proposed adaptive weight PSO-based algorithm and compared with interior point method (IPM) and GA techniques. The suggested technique outperforms both IPM and GA considering exploitation capability. This demonstrates the superiority of bio-inspired algorithm.

Poursistani et al. [45] used an optimization technique founded on binary type of gravitational search algorithm (BGSA) in order to plan the optimal charging of PEVs. The results showed positive effect of smart charging using BGSA technique for the peak shaving of load.

Rahman et al. employed accelerated particle swarm optimization (APSO) [46] and hybrid particle swarm optimization and GSA (PSO-GSA) [47] for SoC maximization of PEVs, hence optimizing the overall smart charging. The hybrid

Table 9.1 A summary of different bio-inspired CI techniques for PEV charging optimization

Authors and references	Applications	Bio-inspired CI technique	Year
Xu et al. [33]	Charge scheduling optimization	ACO	2013
Crow [34]		GA	2014
Soares et al. [35]		PSO	2013
Ghanbarzadeh et al. [36]		ACO, HPSO	2011
Govardhan and Roy [37]		TLBO	2015
Rorigues et al. [44]	Optimal charging strategy	AIS	2013
Bashash et al. [38]		NSGA	2011
Su and Chow [39]		PSO	2012
Poursistani et al. [45]		BGSA	2015
Rahman et al. [46]		APSO	2016
Vasant et al. [47]		PSO-GSA	2016
Awasthi et al. [49]		GA-PSO	2017

PSO-GSA uses the benefits of both GSA and PSO techniques and thus obtains optimum fitness values. Nevertheless, PSO-GSA technique shows much higher computational time compared to single techniques because of complex algorithm formulation. Normally, hybrid bio-inspired methods produce good results for best fitness value. However, the computational time is longer compared to single method [48] because more parameters initialization is needed due to two different methods working in parallel in the hybrid algorithm.

Moreover, Awasthi et al. hybridized GA with an enhanced variant of the particle swarm optimization (GAIPSO) in order to find best location for suggested charging strategy in the Allahabad power distribution company, India [49]. Simulation analysis on a real-time Allahabad city power system clearly shows the superior capability of the stated method compared to PSO and GA to optimize the fitness function considering voltage profile improvement and solution quality.

Table 9.1 summarizes a number of very recent bio-inspired CI techniques applied for the discussed applications of PEV charging optimization.

9.6 Summary and Conclusion

Scientists from multi-disciplinary research backgrounds should try to apply the theoretical knowledge to solve real-time PEV charging problems. Researchers from various backgrounds such as architecture, civil engineering, mechanical engineering and electrical engineering should put painstaking effort together in order to realize successful PEV charging optimization in smart grid. The application of bio-inspired CI techniques for PEV charging optimization is an emerging research field eliciting considerable research attention. The discussions in Sect. 9.5 demonstrate that the overall performances of various bio-inspired CI techniques (specially, PSO, GA) in this domain are very noteworthy as they will inspire other researchers to formulate latest bio-inspired CI techniques to optimize PEV charging. Judging from the trend, interesting variants in this area are hybrid bio-inspired CI such as GA-PSO [47] and PSO-GSA [49] as the results have been found to be highly competitive compared to single CI techniques.

In the future, it is suggested that some well-performed bio-inspired CI techniques like cuckoo search algorithm (CSA), artificial bee colony (ABC) and artificial fish swarm algorithm (AFSA) should be applied to solve issues related to PEV charging. As CSA [50] is constructed on the brood parasitism behavior of cuckoo species and because it uses levy flights, the method can be used for solving PEV complex problems. Levy flight is able to give better result than simple random walk. Moreover, artificial bee colony (ABC) technique is lately presented swarm-based optimization methods which mimics the clever honeybee swarm foraging behavior [51]. It is encouraged that future studies for solving smart charging problem of PEV involve the ABC optimization technique. Furthermore, integrating new or modified hunting behavioral strategies adapted from other bio-inspired algorithms into the swarming behavior stage of AFSA should be able to enhance its convergence rates and optimal solutions of PEV charging [52]. Although various bio-inspired CI techniques have been presented to perform better optimization than that of the standalone versions, the 'No Free Lunch (NFL)' theory [53] is a fundamental barrier toward the overstated claims of the efficiency and robustness of any particular optimization techniques. Particularly, in practice, there is no single optimization technique that can perform best for all types of power system optimization problems. Hence, one potential approach to handle the undesirable implication of the NFL theory is to formulate algorithms based on the synthesis of existing ones as well as limit the applications of a given algorithm to only a specific type of PEV charging optimization problems.

References

1. I. Rahman, P.M. Vasant, B.S.M. Singh, M. Abdullah-Al-Wadud, Intelligent energy allocation strategy for PHEV charging station using gravitational search algorithm, in *AIP Conference Proceedings* (2014), pp. 52–59
2. N. Adnan, S.M. Nordin, I. Rahman, Adoption of PHEV/EV in Malaysia: a critical review on predicting consumer behaviour. Renew. Sustain. Energy Rev. **72**, 849–862 (2017)
3. N. Adnan, S.M. Nordin, I. Rahman, P.M. Vasant, A. Noor, A comprehensive review on theoretical framework-based electric vehicle consumer adoption research. Int. J. Energy Res. (2016)
4. Q. Wang, X. Liu, J. Du, F. Kong, Smart charging for electric vehicles: a survey from the algorithmic perspective. IEEE Commun. Surv Tutorials **18**, 1500–1517 (2016)
5. M.H. Amini, M.P. Moghaddam, O. Karabasoglu, Simultaneous allocation of electric vehicles' parking lots and distributed renewable resources in smart power distribution networks. Sustain. Cities Soc. **28**, 332–342 (2017)

6. H. Shareef, M.M. Islam, A. Mohamed, A review of the stage-of-the-art charging technologies, placement methodologies, and impacts of electric vehicles. Renew. Sustain. Energy Rev. **64**, 403–420 (2016)
7. J. Hu, H. Morais, T. Sousa, M. Lind, Electric vehicle fleet management in smart grids: a review of services, optimization and control aspects. Renew. Sustain. Energy Rev. **56**, 1207–1226 (2016)
8. Z. Yang, K. Li, A. Foley, Computational scheduling methods for integrating plug-in electric vehicles with power systems: a review. Renew. Sustain. Energy Rev. **51**, 396–416 (2015)
9. E.S. Rigas, S.D. Ramchurn, N. Bassiliades, Managing electric vehicles in the smart grid using artificial intelligence: a survey. IEEE Trans. Intell. Trans. Syst. **16**, 1619–1635 (2015)
10. A. Foley, I. Winning, B.Ó. Gallachóir, State-of-the-art in electric vehicle charging infrastructure, in *2010 Vehicle Power and Propulsion Conference (VPPC)* (IEEE, 2010), pp. 1–6
11. F. Mwasilu, J.J. Justo, E.-K. Kim, T.D. Do, J.-W. Jung, Electric vehicles and smart grid interaction: a review on vehicle to grid and renewable energy sources integration. Renew. Sustain. Energy Rev. **34**(6), 501–516 (2014)
12. P. Kulshrestha, L. Wang, M.-Y. Chow, S. Lukic, Intelligent energy management system simulator for PHEVs at municipal parking deck in a smart grid environment, in *2009 Power and Energy Society General Meeting, PES'09.* (IEEE, 2009), pp. 1–6
13. C. Pang, P. Dutta, S. Kim, M. Kezunovic, I. Damnjanovic, PHEVs as dynamically configurable dispersed energy storage for V2B uses in the smart grid, in *IET Conference Proceedings* (2010), pp. 174–174, http://digital-library.theiet.org/content/conferences/10.1049/cp.2010.0903
14. L. Herrera, R. Murawski, F. Guo, E. Inoa, E. Ekici, and J. Wang, PHEVs charging stations, communications, and control simulation in real time, in *Vehicle Power and Propulsion Conference (VPPC)* (IEEE, 2011), pp. 1–5
15. E. Inoa, F. Guo, J. Wang, W. Choi, A full study of a PHEV charging facility based on global optimization and real-time simulation, in *2011 IEEE 8th International Conference on Power Electronics and ECCE Asia (ICPE & ECCE)* (2011), pp. 565–570
16. Z. Ren, H. Jiang, J. Xuan, Z. Luo, Hyper-heuristics with low level parameter adaptation. Evol. Comput. **20**, 189–227 (2012)
17. P. Tulpule, V. Marano, G. Rizzoni, Effects of different PHEV control strategies on vehicle performance, in *2009 American Control Conference, ACC'09* (2009), pp. 3950–3955
18. N.H. Tehrani, G. Shrestha, P. Wang, Optimized power trading of a PEV charging station with energy storage system, in *IPEC* (2012), p. 305
19. F. Pan, R. Bent, A. Berscheid, D. Izraelevitz, Locating PHEV exchange stations in V2G, in *2010 First IEEE International Conference on Smart Grid Communications (SmartGrid-Comm)* (2010), pp. 173–178
20. A. Elgammal, A. Sharaf, Self-regulating particle swarm optimised controller for (photovoltaic-fuel cell) battery charging of hybrid electric vehicles. Electr. Syst. Trans. IET **2**, 77–89 (2012)
21. F. Fazelpour, M. Vafaeipour, O. Rahbari, M.A. Rosen, Intelligent optimization to integrate a plug-in hybrid electric vehicle smart parking lot with renewable energy resources and enhance grid characteristics. Energy Convers. Manag. **77**, 250–261 (2014)
22. W. Su, *Performance Evaluation of an EDA-Based Large-Scale Plug-In Hybrid Electric Vehicle Charging Algorithm* (2012)
23. M.H. Amini, A. Kargarian, O. Karabasoglu, ARIMA-based decoupled time series forecasting of electric vehicle charging demand for stochastic power system operation. Electr. Power Syst. Res. **140**, 378–390 (2016)
24. H.M. Neumann, D. Schär, F. Baumgartner, The potential of photovoltaic carports to cover the energy demand of road passenger transport. Prog. Photovoltaics Res. Appl. **20**, 639–649 (2012)
25. G. Rizzo, I. Arsie, M. Sorrentino, Solar energy for cars: perspectives, opportunities and problems, in *GTAA Meeting* (2010), pp. 1–6

26. P.J. Tulpule, V. Marano, S. Yurkovich, G. Rizzoni, Economic and environmental impacts of a PV powered workplace parking garage charging station. Appl. Energy **108**, 323–332 (2013)
27. D.P. Birnie, Solar-to-vehicle (S2V) systems for powering commuters of the future. J. Power Sources **186**, 539–542 (2009)
28. Q. Zhang, T. Tezuka, K.N. Ishihara, B.C. Mclellan, Integration of PV power into future low-carbon smart electricity systems with EV and HP in Kansai Area, Japan. Renew. Energy **44**, 99–108 (2012)
29. S. Binitha, S.S. Sathya, A survey of bio inspired optimization algorithms. Int. J. Soft Comput. Eng. **2**, 137–151 (2012)
30. M. Črepinšek, S.-H. Liu, M. Mernik, Exploration and exploitation in evolutionary algorithms: a survey. ACM Comput. Surv. (CSUR) **45**, 35 (2013)
31. M. Črepinšek, S.-H. Liu, M. Mernik, Replication and comparison of computational experiments in applied evolutionary computing: common pitfalls and guidelines to avoid them. Appl. Soft Comput. **19**, 161–170 (2014)
32. X.-S. Yang, Z. Cui, R. Xiao, A.H. Gandomi, M. Karamanoglu, *Swarm Intelligence and Bio-Inspired Computation: Theory and Applications* (Newnes, 2013)
33. S. Xu, D. Feng, Z. Yan, L. Zhang, N. Li, L. Jing, et al., Ant-based swarm algorithm for charging coordination of electric vehicles. Int. J. Distrib. Sens. Netw. (2013)
34. M.L. Crow, Economic scheduling of residential plug-in (hybrid) electric vehicle (PHEV) charging. Energies **7**, 1876–1898 (2014)
35. J. Soares, H. Morais, Z. Vale, Particle swarm optimization based approaches to vehicle-to-grid scheduling, in *2012 Power and Energy Society General Meeting* (IEEE, 2012), pp. 1–8
36. T. Ghanbarzadeh, S. Goleijani, M.P. Moghaddam, Reliability constrained unit commitment with electric vehicle to grid using hybrid particle swarm optimization and ant colony optimization, in *2011 Power and Energy Society General Meeting* (IEEE, 2011), pp. 1–7
37. M. Govardhan, R. Roy, Economic analysis of unit commitment with distributed energy resources. Int. J. Electr. Power Energy Syst. **71**, 1–14 (2015)
38. S. Bashash, S.J. Moura, J.C. Forman, H.K. Fathy, Plug-in hybrid electric vehicle charge pattern optimization for energy cost and battery longevity. J. Power Sources **196**, 541–549 (2011)
39. W. Su, M.-Y. Chow, Performance evaluation of a PHEV parking station using particle swarm optimization, in *2011 Power and Energy Society General Meeting* (IEEE, 2011), pp. 1–6
40. I. Fister, D. Strnad, X.-S. Yang, I. Fister Jr, Adaptation and hybridization in nature-inspired algorithms, in *Adaptation and Hybridization in Computational Intelligence* (Springer, 2015), pp. 3–50
41. B. Xing, W.-J. Gao, *Innovative Computational Intelligence: A Rough Guide to 134 Clever Algorithms*, vol. 62 (Springer, 2014)
42. P.-Y. Kong, G.K. Karagiannidis, Charging schemes for plug-in hybrid electric vehicles in smart grid: a survey. IEEE Access **4**, 6846–6875 (2016)
43. I. Rahman, P.M. Vasant, B.S.M. Singh, M. Abdullah-Al-Wadud, Novel metaheuristic optimization strategies for plug-in hybrid electric vehicles: a holistic review. Intell. Decision Technol. **10**, 149–163 (2016)
44. Y.R. Rorigues, M.F. Souza, B. Lopes, A. Souza, D. Oliveira, Recharging process of plug in vehicles by using artificial immune system and tangent vector (2013)
45. M. Poursistani, M. Abedi, N. Hajilu, G. Gharehpetian, Smart charging of plug-in electric vehicle using gravitational search algorithm, in *2014 Smart Grid Conference (SGC)* (2014), pp. 1–7
46. I. Rahman, P.M. Vasant, B.S.M. Singh, M. Abdullah-Al-Wadud, On the performance of accelerated particle swarm optimization for charging plug-in hybrid electric vehicles. Alexandria Eng. J. **55**, 419–426 (2016)
47. P.M. Vasant, I. Rahman, B. Singh Mahinder Singh, M. Abdullah-Al-Wadud, Optimal power allocation scheme for plug-in hybrid electric vehicles using swarm intelligence techniques. Cogent Eng. **3**, 1203083 (2016)

48. T. Ting, X.-S. Yang, S. Cheng, K. Huang, Hybrid metaheuristic algorithms: past, present, and future, in *Recent Advances in Swarm Intelligence and Evolutionary Computation* (Springer, 2015), pp. 71–83
49. A. Awasthi, D. Chandra, S. Rajasekar, A.K. Singh, K.M. Perumal, Optimal infrastructure planning of electric vehicle charging stations using hybrid optimization algorithm, in *2016 Power Systems Conference (NPSC)* (National, 2016), pp. 1–6
50. M. Basu, A. Chowdhury, Cuckoo search algorithm for economic dispatch. Energy **60**, 99–108 (2013)
51. N. Sulaiman, J. Mohamad-Saleh, A.G. Abro, A modified artificial bee colony (JA-ABC) optimization algorithm, in *Proceedings of the International Conference on Applied Mathematics and Computational Methods in Engineering* (2013), pp. 74–79
52. M. Neshat, G. Sepidnam, M. Sargolzaei, A.N. Toosi, Artificial fish swarm algorithm: a survey of the state-of-the-art, hybridization, combinatorial and indicative applications. Artif. Intell. Rev. 1–33 (2014)
53. D.H. Wolpert, W.G. Macready, No free lunch theorems for optimization. IEEE Trans. Evol. Comput. **1**, 67–82 (1997)

Part IV
Promises of Power Grids for Sustainable Interdependent Networks

Chapter 10
Coordinated Management of Residential Loads in Large-Scale Systems

Amir Safdarian

10.1 Residential Demand Response

Smart grid is defined as an evolved electric grid equipped with advanced infrastructures and technologies which enable metering, monitoring, and control of the system. The grid is operated in an affordable, environmentally friendly, reliable, and sustainable manner. These all are made possible by smart grid features like demand response, distributed generation, network automation, renewable energy resources, and self-healing capabilities. Among them, demand response has captured significant attention and great efforts have been dedicated to realize the potentials and benefits of demand response. Demand response as one of the key aspects of future smart grids is a mechanism to motivate consumers, by either time-varying prices or monetary incentives, to adjust their electricity use to improve power system operations. The North American Electric Reliability Corporation (NERC) estimated the contribution of demand response to service reliability issues [1]. In 2012, Federal Energy Regulatory Commission (FERC) declared that demand response has capacity to alleviate US peak demand by at least 9% [1]. The focus of the US government on demand response goes back to the fact that about 25% of US generation and 10% of US transmission and distribution facilities are only necessary during 400 h of annual peak periods [2]. Also, it was estimated that energy efficiency and demand response can lead to 0.4–0.85% incremental reduction in the rate of growing demand in the US [3]. In UK, demand response is claimed to be able to reduce peak demand by more than 15% [4]. In Finland, demand response capacity was estimated to be about 20% of the peak demand [5]. In order to realize the stated potentials, several utilities around the world designed and implemented

A. Safdarian (✉)
Department of Electrical Engineering, Sharif University of Technology, Tehran, Iran
e-mail: safdarian@sharif.edu

© Springer International Publishing AG, part of Springer Nature 2018
M. H. Amini et al. (eds.), *Sustainable Interdependent Networks*,
Studies in Systems, Decision and Control 145,
https://doi.org/10.1007/978-3-319-74412-4_10

sorts of demand response programs. A wide review over the experiences and lessons learnt from demand response programs applied by different US utilities was reported by Berkeley Lab [6]. As the report implies, participation in many of the programs is dominated by large industrial consumers. This is mainly due to the eligibility requirements on consumer size, which restricts smaller consumers from participation. As an instance, a demand response program has been designed in Florida Power and Light (FP and L) where consumers with peak demand higher than 500 kW were allowed to participate. Such eligibility requirements are usually considered in other countries as well. As an example, in the hope of reducing annual peak demand, Ministry of Energy of Iran implemented a demand response program where consumers with peak demand lower than 500 kW were not allowed to participate [7]. This program led to nearly 2% reduction in the national annual peak demand. Although the programs attained great achievements, they overlook the capacity of smaller consumers whose participation may lead to much more significant achievements. In order to apply the capacity of smaller consumers, US government put substantial insistence on the necessity of demand response by residential consumers [8]. This insistence makes sense since more than one-third of worldwide electricity production is consumed by residential sector. Besides the experiences and estimations, several articles focused on potentials, implementations, and enabling technologies of demand response by residential consumers. In [9, 10], distribution systems hosting several residential consumers were studied to demonstrate the potentials. The studies demonstrated that different aspects of distribution system operation can be drastically enhanced via activating demand response. A multi-agent-based system was developed in [11] to enable demand response potentials in a distribution system. In [12–17], system-wide coordination frameworks were developed to achieve most of the benefits and potentials. In [18–20], demand response was integrated in operation of distribution utilities.

Demand response by residential consumers, however, encounters some critical barriers when it comes to real-world implementation [17]. As one of the most important barriers, it is so difficult for small consumers to manually respond to utility signals. In some cases, potential saving in consumers electricity bills is not considerable enough to inspire their active participation in the programs. They do not have enough motivation to continuously follow the signals announced by the utility and respond effectively. The lack of knowledge among small consumers on how to participate in demand response programs is another barrier. They mostly do not have expertise to predict future signals and make the optimal decision effectively. Also, conservative and wealthy consumers do not compromise their comfort to achieve some savings in their electricity bills. The next challenge regards consumers concerns about their privacy. They may believe that the programs may threat their privacy as details of their electricity usage and lifestyle may need to be recorded by the utility. Finally, installing thousands of smart meters that enable effective interaction between the program user and consumers is costly and time-consuming.

In order to overcome the challenges, sorts of solutions have been proposed by industry experts and university researchers. Among the solutions, consumer side systems have captured attention of the industry. The systems mainly mitigate concerns about consumer knowledge on how to participate, consumer lack of expertise in future prediction and decision making, consumer ineffective response to signals, and comfort preservation. The systems usually consist of some sensors, switches, a processing unit, and a communication infrastructure. The sensors are to gather local data on the operation of household devices (e.g., dishwasher, clothes dryer, electric vehicle). These data are necessary to preserve consumer comfort and optimize operation of the devices according to the received signals. Different sensors may be used in a consumer side system. Temperature sensors are used to check if hot water temperature and/or ambient temperature are within the comfortable zone set by the owner. Presence and light sensors are used to check if any lighting bulb can be turned off and if daylight can be used as an alternative for the electric bulbs. The switches are to apply optimal operating plans for the devices. As an example, a switch embedded in the water heater can be used to control on/off operation of the device. The processing unit receives sensors data and utility signals, optimizes operation of the devices, and commands the switches to apply the decisions. This processing unit can be a cheap microcontroller, a personal computer, an application on the owner smart phone, etc. Finally, the communication infrastructure enables interactions between the sensors, switches, and processing unit. Different communication protocols such as WiFi and Zigbee can be used in a consumer side system. In addition, typical consumer side systems may be equipped with some additional features such as a system that access online weather prediction data and a system that receives location data of the home owner from his/her GPS (Global Positioning System) to name just a few. The data provided by these additional features may help to preserve consumer comfort as well as to better optimize operation of the devices.

The consumer side system discussed in the above is the main enabling technology for residential demand response. As mentioned heretofore, the system consists of sensors, switches, processing unit, and communication infrastructure. Among them, processing unit is the most complex part. The unit is equipped with a decision-making process and a procedure for interacting with the program owner. The remainder of this chapter is organized as follows. Section 10.2 focuses on the decision-making process by the processing unit. Sections 10.3 and 10.4 discuss interactions between the processing unit and program owner. Section 10.3 focuses on centralized coordination frameworks where consumers data and preferences are gathered to ensure coordination between consumers responses. Section 10.4 is dedicated to distributed frameworks. The section explains two sequential and non-sequential frameworks. In the distributed frameworks, consumers responses are determined locally and coordinated via system-wide interactions with other consumers and the program owner. Finally, concluding remarks are drawn in Sect. 10.5.

10.2   Home-Wide Management

As described in the preceding section, the processing unit embedded in consumer side system optimizes operation of the home devices. This optimization can be done via different approaches. Meta-heuristic approaches (e.g., genetic algorithm), heuristic and rule of thumb approaches (e.g., greedy method), and mathematical models can be applied to the optimization. Meta-heuristic approaches need sophisticated and time-consuming computations. These approaches do not guarantee the global optimal solution. Also, they suffer from achieving different solutions in successive simulations. Heuristic and rule of thumb approaches do not guarantee the global optimal solution as well. They are usually simple and can be easily applied in cheap off the shelf processors. The mathematical models are, however, able to achieve the global optimal solution. The most complicated part of using these models is to transform the problem into an appropriate model. Recent concerns about the complexity of solving mathematical models are alleviated since processing task can be done via solvers in smart phones and/or online solvers accessible through internet. Putting implementation complexities aside, this chapter focuses on mathematical models which lead to high-quality solutions. The mathematical model is an optimization problem whose objective can be cost minimization, energy saving maximization, or a desired load profile. The objective is subject to some constraints that mainly preserve consumer comfort and operational conditions of the devices. The model needs sorts of data like desired ambient temperature, desired hot water temperature, operational conditions of the devices, and comfort limits set by the owner as input. It also determines operation plan for the devices as output. Mathematical representation of the model is presented in the following [14].

The model can be formulated in the hope of either minimizing cost of electricity procurement, energy saving maximization, or achieving a desired load profile. In many occasions, electricity bill is the only issue concerned by consumers. So, without loss of generality, cost minimization is considered here as follows:

$$\text{Minimize} \quad C = \sum_{t \in T} \rho_t P_t \Delta t \tag{10.1}$$

where t and T are index and set of time intervals. P_t represents demand of the consumer at time t. C is the energy expense of the consumer. $\rho_{n,t}$ is electricity sale price offered at time t. Finally, Δt is the duration of time intervals. The problem is subject to some constraints which are described hereinafter.

The demand of the consumer at time t is calculated as follows:

$$P_t = P_t^f + P_t^{ev} + P_t^{su} + P_t^{hvac} + P_t^{ewh} - P_t^{dg} + \sum_{a \in A} P_{a,t}; \quad \forall t \in T \tag{10.2}$$

where P_t^f is the fixed load of the consumer at time t. P_t^{ev}, P_t^{su}, and P_t^{dg} represent the charging power of electric vehicle, the charging power of storage unit, and the generating power of distributed generation belong to the consumer at time t. a and A are index and set of responsive appliances, respectively. Finally, $P_{a,t}$ is the demand of appliance a at time t. P_t^{hvac} and P_t^{ewh} represent power absorptions of HVAC (heating, ventilation, and air-conditioning) and EWH (electric water heater) loads at time t.

The following constraints are associated with the operation and energy consumption of responsive appliances.

$$\sum_{t \in T} P_{a,t} \Delta t = E_a; \quad \forall a \in A \tag{10.3}$$

$$P_{a,t} = x_{a,t} P_a; \quad \forall a \in A, \forall t \in T \tag{10.4}$$

$$x_{a,t} = 0; \quad \forall a \in A, \forall t \in T - [\alpha_a, \beta_a] \tag{10.5}$$

$$\sum_{t \in T} z_{a,t} = 1; \quad \forall a \in A \tag{10.6}$$

$$\frac{x_{a,t} - x_{a,t-1}}{2} - \varepsilon \leq z_{a,t} \leq 1 + \frac{x_{a,t} - x_{a,t-1}}{2} - \varepsilon; \quad \forall a \in A, \forall t \in T \tag{10.7}$$

where E_a is the total energy required for a complete operation cycle of appliance a. $x_{a,t}$ is a binary variable denoting operating status of appliance a at time t. $x_{a,t}$ is 1 if the associated appliance is operating during the time interval and zero otherwise. P_a is the demand of appliance a when operating. α_a and β_a are, respectively, the beginning and end of allowable operation time for appliance a. $z_{a,t}$ is a binary variable denoting start-up status of appliance a at time t. $z_{a,t}$ is 1 if the associated appliance is started to operate during the time interval and zero otherwise. ε is an arbitrary small number. Equation (10.3) ensures that the energy required for a complete operation cycle of each appliance is provided. Equation (10.4) ensures that the power demanded by each appliance is equal to its nominal power rating if it is energized and 0 otherwise. Equation (10.5) guarantees that all appliances are operated within the period determined by the consumer. This is to ensure that consumer comfort and preferences are intact. Equation (10.6) ensures that once an appliance is energized, its operation will continue to the time when a full operation cycle is achieved. It is worth mentioning that (10.6) is not needed to be adhered for appliances whose operation can be interrupted. As an example, operation of a water heater can be interrupted without encountering any issue. Equation (10.7) is considered to prevent any conflicting situation in binary variables denoting the operation status of responsive appliances. According to (10.7), $z_{a,t}$ is 1 only if $x_{a,t}$ is 1 and $x_{a,t-1}$ is 0, implying that the associated appliance is started up at time t.

It is worthwhile to note that the power required by each appliance during its operating cycle is assumed to be constant. This assumption is, however, not accurate since detailed energy consumption profiles associated with appliances are

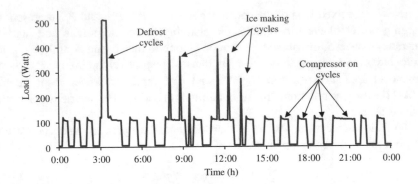

Fig. 10.1 Energy usage profile of a typical refrigerator

ignored. This inaccuracy is usually inevitable mainly due to the fact that the energy consumption profiles of appliances are not available. Also, the profiles usually depend on several factors such as manufacturer, age, and loading. Figure 10.1 depicts energy consumption profile of a typical refrigerator [21]. As can be seen, the refrigerator consumes electricity for compressor, defrost, and ice maker cycles whose occurrence time depends on some out of control factors such as ambient temperature and the volume of stored foods.

The following constraints are taken into account to ensure that sufficient energy is stored in electric vehicle within the time it is parked at home.

$$E_t^{ev} = E_{t-1}^{ev} + P_t^{ev} \Delta t; \quad \forall t \in T \tag{10.8}$$

$$E_t^{ev} \geq E^{ev}; \quad t = \beta^{ev} \tag{10.9}$$

$$P_t^{ev} = 0; \quad \forall t \in T - [\alpha^{ev}, \beta^{ev}] \tag{10.10}$$

$$P_t^{ev} \leq \bar{P}^{ev}; \quad \forall t \in T \tag{10.11}$$

where E_t^{ev} is the total stored energy in battery of electric vehicle at time t. E^{ev} is the total energy required for the trip of the electric vehicle. α^{ev} and β^{ev} are, respectively, arrival and departure times of the electric vehicle trip (i.e., end and beginning times of the consumer trip). $\bar{P}^{ev}$ is the charging rate of electric vehicle battery. Equation (10.8) is used to calculate the total stored energy in the battery. Equation (10.9) ensures that enough energy is stored in electric vehicle battery before its departure time. In other words, this guarantees that consumer comfort and preferences are not sacrificed. Equation (10.10) is taken into account to ensure that the electric vehicle is not charged when it is not at home. Equation (10.11) forces charging power of the electric vehicle under its nominal charging rate.

HVAC load is modeled via a thermodynamic model wherein dynamics of ambient temperature are captured using a lumped capacity representing building fabric heat capacity and air heat capacity [15]. Applying the model, ambient temperature is calculated as follows:

$$C_{buil}\frac{T_t^{ia} - T_{t-1}^{ia}}{\Delta t} = \eta^{hvac}P_t^{hvac} + H^{io}(T_t^{oa} - T_t^{ia}) + H^{ig}(T_t^g - T_t^{ia})$$
$$+ H^{iv}(T_t^v - T_t^{ia}); \quad \forall t \in T \tag{10.12}$$

where C_{buil} denotes building thermal capacity. η^{hvac} is efficiency coefficient of the HVAC system. H^{io}, H^{ig}, and H^{iv} are the thermal conductance between indoor air and outdoor air, the thermal conductance between indoor air and ground, and the thermal conductance between indoor air and ventilation air, respectively. T_t^{ia}, T_t^g, T_t^{oa}, and T_t^v denote indoor ambient air temperature, ground temperature, outdoor ambient air temperature, and ventilation air temperature at time t. The rating capacity of HVAC system is considered as follows:

$$0 \leq P_t^{hvac} \leq \overline{P^{hvac}}; \quad \forall t \in T \tag{10.13}$$

where $\overline{P^{hvac}}$ represents HVAC rating capacity. To ensure that indoor ambient temperature is within the acceptable range, the following constraint is considered.

$$\underline{T^{ia}} \leq T_t^{ia} \leq \overline{T^{ia}}; \quad \forall t \in T \tag{10.14}$$

where $\overline{T^{ia}}$ and $\underline{T^{ia}}$ are upper and lower bounds of acceptable range for indoor temperature. Applying equivalent thermal parameter model, domestic hot water temperature is calculated as follows [15]:

$$C_w(V - W_t)\frac{T_t^{hw} - \eta^{hws}T_{t-1}^{hw}}{\Delta t} = \eta^{ewh}P_t^{ewh} - C_w W_t\frac{T_t^{hw} - T_t^g}{\Delta t}; \quad \forall t \in T \tag{10.15}$$

where C_w is heat capacity of a unit volume of water. V is volume capacity of hot water storage. W_t represents hot water consumption at time t. η^{hws} and η^{ewh}, respectively, denote efficiency coefficient of hot water storage and electric water heater. T_t^{hw} is hot water temperature at time t. The rating capacity of EWH system is considered as follows:

$$0 \leq P_t^{ewh} \leq \overline{P^{ewh}}; \quad \forall t \in T \tag{10.16}$$

where $\overline{P^{ewh}}$ is EWH rating capacity. To ensure that hot water temperature is within the acceptable range, the following constraint is considered:

Table 10.1 Technical data and consumers preferences

Consumer	Item	Responsive appliance			
		Dishwasher	Clothes washer	Clothes dryer	Electric vehicle
#1	Consumption per action (kWh)	1.19	2.47	0.89	–
	Time steps per action	4	7	3	–
	Allowable operation period	10:00–18:30	10:00–21:15	13:00–15:00	–
#2	Consumption per action (kWh)	1.19	2.47	0.89	–
	Time steps per action	4	7	3	–
	Allowable operation period	14:30–19:45	12:30–14:15	11:45–14:15	–
#3	Consumption per action (kWh)	1.19	2.47	0.89	7.33
	Time steps per action	4	7	3	4
	Allowable operation period	17:15–22:00	15:30–19:00	10:15–13:00	20:00–9:00
#4	Consumption per action (kWh)	1.19	2.47	0.89	–
	Time steps per action	4	7	3	–
	Allowable operation period	11:30–17:45	7:15–17:45	12:00–12:30	–

$$\underline{T^{hw}} \leq T_t^{hw} \leq \overline{T^{hw}}; \quad \forall t \in T \tag{10.17}$$

where $\overline{T^{hw}}$ and $\underline{T^{hw}}$ are upper and lower bounds of acceptable range for hot water temperature.

In the above problem, the operating status of different responsive appliances and operating power of HVAC and EWH systems are independent decision variables of the model.

The above model is applied to the load profile of 4 residential consumers. The consumers are assumed to have three responsive appliances, i.e., dishwashers, clothes washers, and clothes dryers. The third consumer has electric vehicle as well. The simulations are conducted with the time resolution of 15 min. Table 10.1 provides technical data and consumer preferences associated with the responsive appliances.

Individual consumers load profiles before any response to the signals received from the system operator are given in Fig. 10.2. Here, it is assumed that the responsive appliances are operated at the beginning of the associated allowable operation periods. This can be translated to the fact that no delay in the operation of the appliances is applied by the consumers.

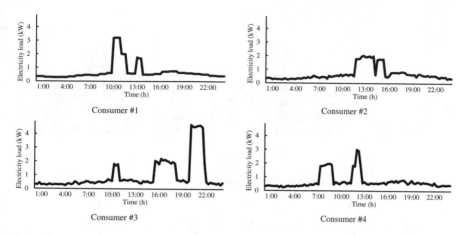

Fig. 10.2 Individual consumers load profiles before responding to system operator signals

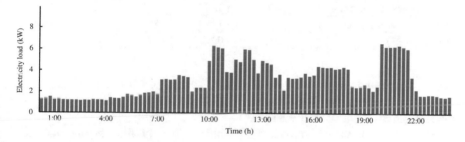

Fig. 10.3 System load profile before responding to system operator signals

Summing up individual consumers load profiles, system load profile is achieved and depicted in Fig. 10.3. As can be seen, system peak load happens at 8 p.m. when total system demand is about 6.6 kW. Also, system load is higher during morning and evening periods while the least demand occurs around midnight from 10 p.m. to 7 a.m. The system load during off-peak period is less than 2 kW which is about 3 times lower than the peak load.

Here, time-varying prices are announced to the consumers as signals from the system operator. Figure 10.4 depicts the prices which are used in the simulations. As can be observed, electricity prices are higher during morning and evening periods while the afternoon and midnight are the least expensive periods. The price changes motivate the consumers to defer the operation of their responsive appliances from morning to noon and from evening to midnight.

Responding to the price signals, consumers energy expenses are decreased. Energy expenses of the consumers before and after responding to the prices are given in Table 10.2. As it was expected, energy expenses of all consumers decrease as they respond to the announced prices. The decrement ranges from less than 1% for the second consumer to more than 4% for the third consumer. According to the

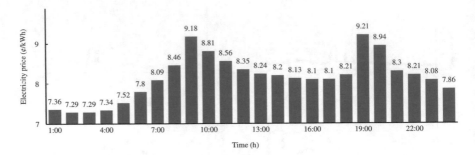

Fig. 10.4 Hourly electricity prices as system operator signals

Table 10.2 Consumers energy expenses before and after responding to system operator signals

Energy expense	Consumer				Total
	#1	#2	#3	#4	
Before consumers response ($)	1.377	1.375	1.978	1.387	6.117
After consumers response ($)	1.368	1.374	1.893	1.384	6.019
Reduction (%)	0.63	0.10	4.47	0.24	1.63

results, total energy expense of the system decreases by 9.8 cents which is about 1.6% of the total expense. The electricity price has the highest value at 7 p.m., while the least prices occur around 3 a.m.

Figure 10.5 compares individual consumers load profiles before and after responding to the prices. As can be seen, the second and third consumers, respectively, experience the least and most changes in the associated load profiles. This results in the least and most significant savings in energy expenses of the second and third consumers, respectively. Albeit, it should be considered that the third consumer has an electric vehicle which is a responsive appliance with considerable energy use and flexibility.

Figure 10.6 compares total system load profiles before and after consumers response to the price signals. As can be seen, significant portion of system load during peak periods (i.e., from 8 p.m. to 9:30 p.m.) is postponed to off-peak periods (i.e., from 0:45 a.m. to 2:15 a.m.). Total system peak load is displaced from 8 p.m. to 3:30 p.m. This displacement is accompanied with about 0.7 kW increment in value. The peak load is increased from 6.6 kW by about 11% to 7.33 kW. This increment in system peak load, in spite of prevalent believes about incontestable benefits of implementing demand response, may deteriorate system efficiency. It is worthwhile to mention that peak rebounds like what observed in the simulations are likely when consumers respond to price signals autonomously.

The above observations imply that autonomous response by individual consumers, although it is beneficial for the consumers themselves, can worse system operational conditions. In order to prevent the negative effects, system-wide coordination approaches are necessary.

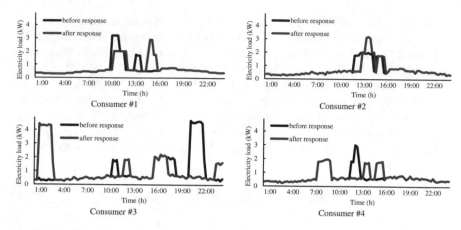

Fig. 10.5 Individual consumers load profiles before and after responding to system operator signals

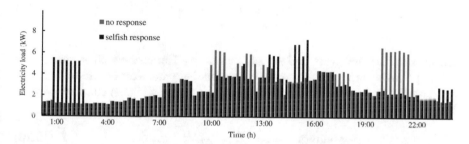

Fig. 10.6 System load profile before and after responding to system operator signals

10.3   Centralized System-Wide Coordination

Heretofore, mathematical model for optimal response of residential consumers to signals (e.g., hourly electricity prices) received from the system operator was described. According to the model, each consumer optimizes operation of his/her responsive appliances such that the associated energy expense is minimized. This, although is beneficial for individual consumers, may threaten electric system efficiency and performance. Actually, selfish actions by individual consumers may lead to new and even worse peak rebounds during off-peak periods when electricity prices are lower. These statements are already justified via the simulations conducted in the preceding subsection and the associated results. This issue accentuates the existence of a system-wide coordination approach that ensures consumers responses are in line with system efficiency. In order to achieve the goal, some system-wide coordination approaches have been developed in the literature. A centralized approach is presented here [14]. The approach is based on a management model which is in charge of coordinating responses of individual consumers. Since consumers are independent self-interested entities, their objectives and constraints must be satisfied in the management model. This condition can be adhered by a bi-level problem wherein individual consumers decision models are nested within the constraints. In the bi-level problem, the upper subproblem targets system efficiency. The responses by individual consumers are modeled via the lower subproblems. Different objectives can be considered for the upper subproblem. Here, without loss of generality, minimizing system peak load is considered as objective to prevent severe peak rebounds, as one of the most likely consequences of non-coordinated response by consumers. This objective is mathematically formulated as follows:

$$\text{Minimize} \quad Peak \geq P_t^{total}; \quad \forall t \in T \tag{10.18}$$

where $Peak$ is the peak load of the system and P_t^{total} is the total load of the system at time t. P_t^{total} can be calculated as follows:

$$P_t^{total} = \sum_{n \in N} P_{n,t}; \quad \forall t \in T \tag{10.19}$$

where n and N are index and set of consumers. Hereinafter, all symbols with subscript n are used to represent the corresponding parameter but associated with consumer n. As described in the preceding subsection, individual consumers would like to minimize their electricity expenses as follows:

$$\text{Minimize} \quad C_n = \sum_{t \in T} \rho_t P_{n,t} \Delta t; \quad \forall n \in N \tag{10.20}$$

All constrains that ensure consumers convenience and operation limits of appliances must be considered as follows:

$$\text{Constraints} \ (10.2) - (10.17); \quad \forall n \in N \tag{10.21}$$

It is worth mentioning that (10.21) consists of all constraints declared in (10.2)–(10.17) which are written for all individual consumers.

In the above problem, the operating status of all responsive appliances and the power consumed by all HVAC and EWH systems corresponding to all individual consumers are independent decision variables of the model. Once the problem is solved, operation status of all of the devices is determined such that consumers energy expenses are minimized and severe peak rebounds are avoided. However, bi-level optimization problems are very difficult to solve. In many occasions, there is no general step-by-step approach that guarantees achieving to the global optimal solution. The described bi-level model is fortunately a special class of bi-level programs that can be relaxed to an equivalent single-level problem. In the model, solution of the upper subproblem is independent of the objective of the lower subproblems. This implies that the optimum value of the objective in lower subproblems. This implies that the optimum value of the objective in lower sub-problems is not affected by the upper subproblem. Hence, the lower subproblems can be solved independently and their obtained optimum values are then used to relax (10.20) to a set of constraints. To do so, the following problems are firstly solved:

$$\text{Minimize} \quad C_n = \sum_{t \in T} \rho_t P_{n,t} \Delta t \qquad ; \quad \forall n \in N \qquad (10.22)$$

$$\text{subject to} \quad \text{Constraints } (10.2) - (10.17)$$

Assuming that C_n^{opt} is the optimal value of C_n obtained via solving the above problems, (10.20) can be replaced with the following constraints:

$$C_n \leq C_n^{opt}; \quad \forall n \in N \qquad (10.23)$$

$$C_n = \sum_{t \in T} \rho_t P_{n,t} \Delta t; \quad \forall n \in N \qquad (10.24)$$

Now, the bi-level problem can be relaxed to the following single-level problem.

$$\text{Minimize} \quad Peak \geq P_t^{total}; \quad \forall t \in T$$
$$\text{subject to} \quad \text{Constraints } (10.2) - (10.17), (10.19), (10.23) - (10.24) \qquad (10.25)$$

This problem is then solved in a centralized manner. To do so, all input data are gathered in the control center where status of all responsive appliances and power consumed by all HVAC and EWH systems are optimized and sent to the associated consumers. This way, it is guaranteed that system efficiency is maximized as far as no extra expense is imposed on consumers.

The achieved single-level model is applied to the system consisting of the four consumers. Figure 10.7 compares the achieved system load profile with the profiles achieved in the preceding subsections. As can be seen, coordinating the consumers responses via the model leads to significant reduction in peak rebound during off-peak periods. The system peak load is 5.797 kW and occurs at 1:30 p.m. when consumers responses are coordinated.

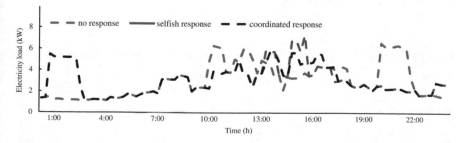

Fig. 10.7 System load profile before and after coordinated and non-coordinated responding to system operator signals

Table 10.3 System peak load before and after coordinated and non-coordinated responding to system operator signals

No response (kW)	Selfish response (kW)	Coordinated response (kW)
6.6	7.331	5.797

Table 10.3 compares system peak load in different situations. According to the results, system peak load experiences about 21% enhancement when the responses are coordinated compared to the situation where the responses are selfish. Also, it experiences more than 12% improvement when coordinated responses are applied compared to the case before consumers respond to the hourly prices. It is worth mentioning that the system efficiency achieved via applying the system-wide coordination approach is without violating consumers preferences and/or imposing extra energy expenses to the consumers.

10.4 Distributed System-Wide Coordination

The preceding section described mathematical model for coordinating responses by residential consumers. The model is then solved in a centralized manner where all required data including technical data of appliances, buildings, and storages as well as consumers preferences are gathered and synthesized in a control center. Although the model has benefits, the centralized fashion may cause a barrier in enrollment of privacy conscious consumers since they may prefer not to announce their preferences and habits. Moreover, the huge data exchange needed in the centralized fashion requires higher capacity communication infrastructures and database systems. In addition, size of the centralized problem dramatically grows as number of consumers increases. This may make the problem very difficult to solve in the real-world systems which host thousands consumers. This huge burden can crash the processors in the control center as well. Table 10.4 demonstrates how size of the problem grows as the number of consumers increases. Here, time resolution and number of responsive appliances are assumed to be 15 min and 3, respectively. As can be observed, the problem would be extraordinarily large in a system hosting several thousands of consumers.

Table 10.4 Size of centralized coordination problem versus the number of consumers

Parameter		Number of consumers				
		1	4	40	400	4000
Number of variables	Binary	576	2304	23040	230400	2304000
	Continuous	1058	3941	38537	384497	3844097
Number of constraints		2697	10212	100392	1002192	10020192

To cope with the pointed challenges, some distributed approaches for solving the coordination problem were presented in the literature. Among different approaches, iterative approaches have captured significant attention. So, two sequential and non-sequential iterative approaches are described in this subsection [14, 16]. The two approaches are described hereinafter.

10.4.1   Sequential System-Wide Coordination Approach

In this approach, the first step is to achieve the minimum energy expenses and load profile of individual consumers. Figure 10.8 depicts the process in this step [14]. As can be seen, to achieve the energy expenses and load profiles, the system operator announces electricity prices. The consumers modify their load profiles such that their energy expenses are minimized. This is done via solving (10.22). Then, the consumers record their minimum energy expenses and provide the operator with the achieved load profiles. Needless to mention, the energy expense minimization by individual consumers may lead to severe peak rebounds at periods with lower electricity prices.

In order to alleviate the peak rebounds, the second step is to sequentially ask consumers to modify their load profiles in such a way that system peak load is diminished while their expenses and comfort remain intact. The second step of the approach is depicted in Fig. 10.9 [14].

At the end of the first step, individual consumers load profiles and energy expenses are achieved and recorded. The consumers send their load profiles to the system operator. The operator simply calculates system load profile by aggregating the received load profiles. He/she sends the achieved system load profile to the first

Fig. 10.8 Sequential coordination approach—the first step

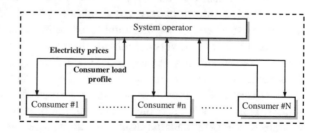

Fig. 10.9 Sequential coordination approach—the second step

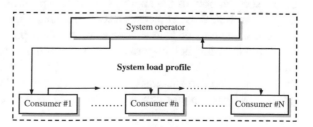

consumer. The consumer modifies his/her load profile such that system peak load is minimized. Then, the consumer sends the modified system load profile to the second consumer. The second consumer tries to diminish system peak load and sends the modified load profile to the next consumer. This process iterates to the time the last consumer minimizes system peak load. Then, the last consumer provides the system operator with the modified load profile. The operator checks if termination criteria are satisfied, and the final system load profile is achieved. Otherwise, the process is restarted by the actions taken by consumers one by one. Usually, the process continues to the point either the number of iterations reaches to a given number or no significant change in system peak load is observed. In order to modify system load profile, consumer n solves the following optimization problem over his/her own decision variables:

$$\text{Minimize} \quad Peak \geq P_t^{total}; \quad \forall t \in T \tag{10.26}$$

$$\text{Constrants} \ (10.2) - (10.17) \tag{10.27}$$

$$P_t^{total} = \sum_{n \in N} P_{n,t}; \quad \forall t \in T \tag{10.28}$$

$$C_n \leq C_n^{opt} \tag{10.29}$$

$$C_n = \sum_{t \in T} \rho_t P_{n,t} \Delta t \tag{10.30}$$

In the above problem, $P_t^{total} - P_{n,t}$ is constant equal to system load minus the load of the consumer at time t. In the distributed approach, total load profile is the only data that consumers send to the system operator. They just have access to the system load profile. So, consumers privacy is adhered. In addition, volume of data exchange is limited to an array of load profile which requires much cheaper communication infrastructures and database systems. Finally, size of the above problem is independent of the number of consumers. This implies the fact that processor needed to solve the problem does not change with the number of consumers. These all may reduce implementation costs and remove barriers in enrollment of consumers in demand response programs.

The distributed approach is applied to the system consisting of the four consumers. The system load profiles when consumers responses are coordinated via the centralized and the sequential distributed approaches are depicted in Fig. 10.10. As can be seen, in both of the approaches, system peak load is equal to 5.797 kW and happens at 1:30 p.m. The two profiles are almost the same. The only differences regard the amounts of load at 3:45 p.m. and 4:45 p.m. It is worth mentioning that the differences are not important since they do not affect the system peak load which is the objective of the two approaches. This implies that the sequential distributed approach successfully coordinates actions taken by the consumers and guarantees system performance and efficiency. Albeit, it should be emphasized that

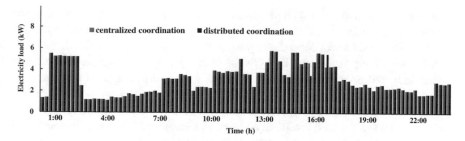

Fig. 10.10 System load profiles after centralized and distributed coordinations of consumers responses

the distributed approach does not guarantee achieving the global optimal solution since the subproblems contain binary variables which make the problem non-convex.

10.4.2   Non-sequential System-Wide Coordination Approach

The sequential distributed approach described in the preceding subsection suffers from long time delay needed for the sequential actions by different consumers. Actually, the sequential decision making of the consumers is necessary to guarantee convergence of the process. In order to mitigate the concerns about the delay, a non-sequential approach is presented in this subsection. In this approach, the first step is to determine consumers selfish responses such that their individual energy expenses are minimized. This is done via solving (10.22). It is clear that the first step in the non-sequential approach is the same as that in the sequential approach. At the end of this step, consumers load profiles and their minimum energy expenses are determined. Then, the system operator gathers the individual load profiles and determines the system load profile by summing up the gathered profiles. It is clear that the achieved load profile is not necessarily in line with system efficiency and performance and coordination in the second step is necessary. The second step of the approach is depicted in Fig. 10.11 [16].

In the second step, the system load profile is announced to the consumers. They non-sequentially investigate if any modification proposal can be provided that

Fig. 10.11 Non-sequential coordination approach—the second step

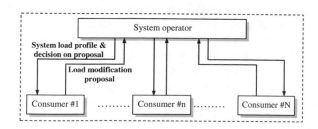

diminish system peak load without violating their minimum energy expenses and comfort. This investigation is conducted via solving (10.26) subject to (10.27)–(10.30). The consumers who have modification proposal provide the system operator with their proposals. The system operator judges the received proposals sequentially and accepts those which enhance system efficiency and performance. The operator sends an acceptance signal to consumers whose proposal is in line with system efficiency. Since system load profile is altered by accepting some proposals, the system load profile is updated and announced to the consumers who would like to provide new proposals. The process is iterated to the time no new proposal is received from the consumers. This way, the time delay is not a concern since consumers generate their proposals non-sequentially. The convergence is also guaranteed since investigating received proposals is conducted in a sequential manner. It is worth mentioning that the investigation is hundred times faster than proposal preparation by consumers. The approach is applied to the four-consumer system which was studied in the preceding subsections. The achieved results are the same as those achieved via the sequential approach. This approach successfully coordinates the response by the consumers and preserves system performance and efficiency. Like the sequential approach, the non-sequential distributed approach does not ensure achieving the global optimal solution since the subproblems contain binary variables which make the problem non-convex. Actually, superiority of the non-sequential approach goes back to the lower time delay that takes to achieve the solution. Simulating the four-consumer system, both of the two approaches need two iterations. This means that the non-sequential approach is about 4 times faster than the sequential approach.

10.5 Conclusion

This chapter first provided brief explanations on demand response by residential consumers and potential barriers that may be encountered during implementation. Lack of knowledge among small consumers on how to participate, lack of expertise among small consumers on how to predict future signals, lack of motivation for continuously following utility signals and manually responding, and concerns about their privacy and comfort are among the key barriers. In order to cope with some of the barriers, automated home-wide management systems were developed. The mathematical model for decision-making process in an automated home-wide management system is formulated in Sect. 10.2. The model was applied to the data of four typical consumers who are provided with hourly electricity prices. Owing to the results, a few percent savings in consumers electricity bills can be achieved when the model is applied. It is worth mentioning that the savings are realized without endangering consumers preferences and comfort. The autonomous response by the consumers, although it reduces individual electricity bills, may worsen system performance and efficiency. According to the results, the autonomous responses led to a severe peak rebound in system load profile during periods with lower electricity prices. In order for system performance and efficiency to remain intact, centralized system-wide coordination approach is described in Sect. 10.3. The model was presented and applied to the four-consumer system. According to the results, severe peak rebounds can be avoided via system-wide coordination approaches. However, the centralized model needs technical data of individual consumers appliances and their preferences to be gathered in a control center. This may lead to several issues such as congestion in communication infrastructures, crash in control center processors, and threatening consumers privacy. To cope with the issues of centralized framework, Sect. 10.4 presents distributed system-wide coordination approaches wherein gathering consumers data is not necessary. The section contains two different approaches, namely sequential and non-sequential approaches. The sequential approach is slower but simpler for real-world implementation. The two approaches were applied to the four-consumer system. The results demonstrated that both of the approaches can successfully mitigate concerns raised by autonomous responses by residential consumers. In brief, the mentioned observations are summarized as follows:

- Applying decision-making process in an automated home-wide system may decrease consumers individual electricity bill. It, however, can lead to severe peak rebounds during periods with lower electricity prices.
- Severe peak rebounds caused by autonomous response by individual consumers can be avoided via coordinating consumers responses.
- Centralized system-wide coordination frameworks effectively coordinate responses by consumers, thereby alleviating peak rebounds. They, however, face implementation complexities especially when number of enrolled consumers grows.
- Distributed system-wide coordination frameworks are effective in coordinating consumers responses. They are fairly simple in implementation as well. However, distributed frameworks hardly guarantee global optimal solution.
- All distributed frameworks have almost identical performance. Non-sequential frameworks are, however, much faster than sequential frameworks. This may ease implementation of non-sequential frameworks in the real world.

Although the literature on residential demand response is rich, future works are headed in several directions as follows:

- System-wide coordination frameworks proposed in the literature are effective in enhancing total system load profile. They, however, do not consider network constraints. Future works can be focused on coordination frameworks which are capable to consider network constraints, thereby enhancing network voltage profiles, network losses, and service reliability.
- Distributed frameworks do not ensure global optimal coordination among consumers responses. So, future works may focus on proposing frameworks that guarantee the global optimal solution. To do so, decomposition algorithms such as Dantzig–Wolfe can be potential approaches.

- Neither of the distributed frameworks is effective in considering the discrete nature of operation modes in appliances such as clothes washer. So, future works can be headed toward developing frameworks which are able to consider the discrete nature. In order to achieve the goal, decomposition algorithms such as Benders decomposition are options.
- In the industry, several smart devices such as smart thermostats have been developed and being used by some consumers. Academic research, however, does not consider the devices and their capabilities. Since academic research and industry actions cannot be developed independently, integrating the smart devices into the models developed by academic research and trying to achieve most of their potentials is an option for future works.
- Many theoretical approaches and frameworks have been developed in the literature. All of the works demonstrated the significant potential of residential demand response and developed approaches to realize the potentials. Now, it is time to implement the approaches in the real world and share practical observations and experiences.
- Proliferation of distributed generation units, electric vehicles, and energy storage units in residential houses is changing the behavior of consumers. Integrating these new technologies in home-wide models and system-wide coordination frameworks is an option for future works.

References

1. U.S. Federal Energy Regulatory Commission, Assessment of Demand Response & Advanced Metering, USA (2008)
2. T.J. Lui, W. Stirling, H.O. Marcy, Get smart. IEEE Power Energy Mag. **8**(3), 66–78 (2010)
3. Electric Power Research Institute, Assessment of achievable potential from energy efficiency and demand response programs in the U.S. (2010–2030), USA (2009)
4. K. Samarakoon, J. Ekanayake, N. Jenkins, Reporting available demand response. IEEE Trans. Smart Grid **4**(4), 1842–1851 (2013)
5. J. Aghaei, M. Alizadeh, Demand response in smart electricity grids equipped with renewable energy sources: a review. Renew. Sustain. Energy Rev. **18**, 64–72 (2013)
6. Ernest Orlando Lawrence Berkeley National Laboratory, A survey of utility experience with real time pricing, Berkeley, CA (2004)
7. Iran Grid Management Company, Findings of demand response by industries, Tehran, Iran (2015)
8. Energy Policy Act of 2005, U.S. Congress, USA (2005)
9. A. Safdarian, M. Fotuhi-Firuzabad, M. Lehtonen, Benefits of demand response on operation of distribution networks: a case study. IEEE Syst. J. **10**(1), 189–197 (2016)
10. A. Safdarian, M.Z. Degefa, M. Lehtonen, M. Fotuhi-Firuzabad, Distribution network reliability improvements in presence of demand response. IET Gener. Transm. Distrib. **8**(12), 2027–2035 (2014)
11. M.H. Amini, B. Nabi, M.R. Haghifam, Load management using multi-agent systems in smart distribution network. IEEE PES General Meeting, Vancouver, British Columbia, Canada (2013)

12. S. Bahrami, M.H. Amini, A decentralized framework for real-time energy trading in distribution networks with load and generation uncertainty. arXiv preprint (2017)
13. M.H. Amini, J. Frye, M.D. Ilić, O. Karabasoglu, Smart residential energy scheduling utilizing two stage mixed integer linear programming, in *IEEE 47th North American Power Symposium*, Charlotte, NC, USA (2015)
14. A. Safdarian, M. Fotuhi-Firuzabad, M. Lehtonen, A distributed algorithm for managing residential demand response in smart grids. IEEE Trans. Ind. Inf. **10**(4), 2385–2393 (2014)
15. A. Safdarian, M. Ali, M. Fotuhi-Firuzabad et al., Domestic EWH and HVAC management in smart grids: potential benefits and realization. Electr. Power Syst. Res. **134**, 38–46 (2016)
16. A. Safdarian, M. Fotuhi-Firuzabad, M. Lehtonen, Optimal residential load management in smart grids: a decentralized framework. IEEE Trans. Smart Grid **7**(4), 1836–1845 (2016)
17. A.H. Mohsenian-Rad, A. Leon-Garcia, Optimal residential load control with price prediction in real-time electricity pricing environment. IEEE Trans. Smart Grid **1**(2), 120–133 (2010)
18. A. Safdarian, M. Fotuhi-Firuzabad, M. Lehtonen, Integration of price-based demand response in DisCos' short-term decision model. IEEE Trans. Smart Grid **5**(5), 2235–2245 (2014)
19. A. Safdarian, M. Fotuhi-Firuzabad, M. Lehtonen, A medium-term decision model for DisCos: forward contracting and TOU pricing. IEEE Trans. Power Syst. **30**(3), 1143–1154 (2015)
20. A. Safdarian, M. Fotuhi-Firuzabad, M. Lehtonen, Impacts of time-varying electricity rates on forward contract scheduling of DisCos. IEEE Trans. Power Deliv. **29**(2), 733–741 (2014)
21. A. Safdarian, Optimal design of price based demand response programs in distribution networks. Dissertation, Sharif University of Technology (2014)

Chapter 11
Estimation of Large-Scale Solar Rooftop PV Potential for Smart Grid Integration: A Methodological Review

Dan Assouline, Nahid Mohajeri and Jean-Louis Scartezzini

11.1 Introduction

In urban areas, decentralized energy systems from nondispatchable energy resources (e.g., wind and solar energy), fully clean and widely available, are progressively replacing traditional power plants. Solar photovoltaics (PV) is a particularly attractive form of renewable generation that can be easily installed on the existing rooftops. The technological developments and permanent cost reduction of PV panels lead to its large-scale deployment in several countries. The optimized use of decentralized generated energy, however, requires the estimation of both supply (e.g., solar PV rooftop) and electricity demand values, along with the design of smart distribution systems and transmission grids. The estimation of solar PV electricity production, in particular at regional and national scale, is crucial for developing the global energy strategies. While estimating rooftop solar PV potential is essential for the grid integration, it is a multifaceted process and requires a rigorous method to do so.

In order to extract the potential for a renewable energy, it is convenient to follow a general approach that divides the extraction process into multiple steps. A hierarchical approach to the energy potential estimation has been proposed by van Wijk and Coelingh [115] and became widely used for many countries and at various scales [55, 60, 120]. The hierarchical approach consists of multiple steps: (i) The *theoretical/physical potential*, which is the theoretical limit of the considered

D. Assouline (✉) · N. Mohajeri · J.-L. Scartezzini
LESO-PB, Ecole Polytechnique Federale de Lausanne, Lausanne, Switzerland
e-mail: dan.assouline@epfl.ch

J.-L. Scartezzini
e-mail: jean-louis.scartezzini@epfl.ch

N. Mohajeri
Sustainable Urban Development Programme, Department for Continuing Education, University of Oxford, Rewley House, 1 Wellington Square, Oxford OX12JA, UK
e-mail: nahid.mohajeri@epfl.ch

© Springer International Publishing AG, part of Springer Nature 2018
M. H. Amini et al. (eds.), *Sustainable Interdependent Networks*,
Studies in Systems, Decision and Control 145,
https://doi.org/10.1007/978-3-319-74412-4_11

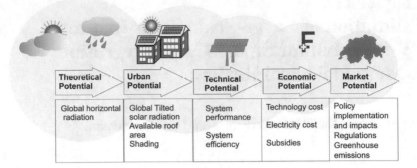

Fig. 11.1 Hierarchical potential approach

resource, meaning the total energy received from the sun at designated locations, (ii) the *geographic/urban potential*, which constraints the theoretical potential over locations where the solar energy can be captured and used for PV installations, in our case the available area over rooftops, (iii) the *technical potential*, which translates into the electrical energy output obtained from the system, considering the various induced losses and the technical characteristics of the PV technology, (iv) the *economic potential*, which accounts for cost constrains involved in the installation of PV, and (v) the *market potential*, which finally restricts the potential according to societal constrains and regulations, including implementation policies, social acceptance, and legal considerations. A scheme of the hierarchical approach principle is showed in Fig. 11.1. It should be acknowledged that while the economic and market potentials may be significantly smaller than the technical potential, they fall out of the scope of this review, and we will not present methods to account for these two last steps.

In the hierarchical approach steps, the difficulty resides in the estimation of a few critical variables. These include mainly: (1) the horizontal global, diffuse, direct, and extraterrestrial solar radiation, (2) the shadowing effects over rooftops, (3) the rooftop slope and aspect distributions, (4) the tilted radiation over rooftops, and (5) the available rooftop area for PV installation. These variables are functions of many different parameters and can be very hard to estimate at a large scale. Depending on the available data, urban characteristics, and scale of the study, various methods have been tested to estimate these variables.

The present chapter aims to provide a review of different methods used in the literature to estimate the variables of interest and finally the solar rooftop PV potential at a regional or national (country) scale. The structure of the rest of the chapter will be as follows: We will present in separate sections different methods leading to a potential estimation, namely (1) physical and empirical models, (2) geostatistical methods, (3) constant-value methods, (4) sampling methods, (5) GIS and LiDAR methods, and (6) machine learning methods, and finally present the summary of the applicability of the methods and their advantages and disadvantages.

In each section, we will first present the principle of the method, along with a theoretical background (if applicable), and then provide a literature review of significant studies applying that method.

11.2 Physical and Empirical Models

The first family of methods historically used for solar potential estimation is the family of physical and empirical models. Physical models refer to analytical formulas derived based on fundamental physical laws. When physical models are available, along with all the variables of interest, they should be used preferably, since they reflect the physical reality of the situation. Yet, in case of very complex and stochastic systems, physical models are troublesome to derive. Also, when such models exist, it is hard to compute them. Therefore, a reasonable alternative is to derive some approximated formulas, based on experience and experimentation, to model a particular problem at hand. Such a model is called an empirical model, since it is validated experimentally and does not rely on theory. One practical advantage of empirical models is that the required computation power can be quite low, depending on the simplicity of the model. Empirical models often include parameters to choose based on local data. Once the parameters are chosen accordingly to the location, the output is obtained simply by plugging in the variables.

In solar potential estimation, physical and empirical models are mainly useful to derive expressions for the different components of the solar radiation, in Wh/m² (also called irradiance). The solar radiation can be split into three different components: direct (or beam), diffuse, and reflected radiations. The sum of the three components is called the global solar radiation. An important point is the inclination of the surface on which the solar radiation is computed. Solar radiation is often measured on a horizontal plane and can be separated into the horizontal direct, diffuse, and reflected solar radiations. In case of a tilted plane (e.g., a typical rooftop), the three components of the radiation need to be re-computed to account for the tilt. They are simply called tilted direct (or beam), diffuse, and reflected radiations, and sum to the global tilted radiation. Since it can be troublesome to directly measure tilted components of the solar radiation, various models exist to derive them using the horizontal solar components values. Models for both tilted and horizontal radiations will be presented further in the section. Notations will be used for the different components of the solar radiation throughout the rest of the section to avoid repetition. They will be as follows:

- G_h, G_B, and G_D are the global horizontal, direct horizontal, and diffuse horizontal radiations.
- G_t, G_{Bt}, and G_{Dt} are the global tilted, direct tilted, and diffuse tilted radiations.
- G_{Bn} is the direct normal radiation, sometimes measured instead of the direct horizontal radiation.

- G_{oh} and G_{on} are the extraterrestrial horizontal and normal radiations, also called the top of atmosphere radiation. It is the radiation coming from the sun before it reaches the earth's atmosphere.
- G_{Rt} is the ground-reflected radiation over a tilted plane, called the reflected tilted radiation.

In this section, we first present the classical physical and empirical models to compute the horizontal radiations and then further present some of the models developed for the components of the tilted radiation. We finally go through the application of solar horizontal radiation and solar tilted radiation models.

11.2.1 Solar Global Horizontal Models

As explained earlier, the horizontal radiation is the sum of the horizontal direct, diffuse, and reflected radiations. The reflected radiation, however, is very negligible over a horizontal plane, compared to the direct and diffuse radiations. Therefore, it is often discarded in the global horizontal radiation that is given by the sum of direct and diffuse horizontal radiations:

$$G_h = G_B + G_D = G_{Bn} \cos(\theta_z) + G_D \qquad (11.1)$$

where θ_z is the solar zenith angle, the angle between the zenith (vertical direction) and the sun, indicating the position of the sun. The smaller the solar zenith angle, the higher the sun is in the sky. As the sun rises, the angle gradually decreases until midday.

To obtain values for G_h, measurements are preferred to any other kind of model or approximation. However, measurement data is not always possible to create. Consequently, parametric empirical models were developed to estimate G_h based on multiple variables that may be available, including extraterrestrial radiation, ambient temperature, shining hours and relative humidity.

The first family of empirical models for global horizontal radiation is the linear family, among which the most famous is the one developed by Angstrom [7]. It expresses the ratio between G_h and G_{oh} as follows:

$$\frac{G_h}{G_{oh}} = a + b \frac{l_d}{S} \qquad (11.2)$$

where l_d and S are, respectively, the day length and number of shining hours. The model parameters a and b are then calculated by fitting the model on some solar radiation data at the location of interest. This model is one of the first empirical models developed. Many more complex linear models were used in the literature, as presented, for instance, in [39].

A variety of nonlinear models can also be found in the literature for global solar radiation: polynomial, logarithmic, exponential, etc. However, one commonly used is the quadratic model [12], which simply adds a second-order nonlinear term to the previous linear model:

$$\frac{G_h}{G_{oh}} = a + b\frac{l_d}{S} + c\left(\frac{l_d}{S}\right)^2 \tag{11.3}$$

where the variables are defined as in the previous model. Other more sophisticated models add terms to account for ambient temperature or relative humidity.

Since it is more delicate to measure diffuse and direct horizontal radiations than it is to measure global horizontal radiation, models have been developed through the years to separate the diffuse and the direct radiations from global radiation measurements. Some of these models were derived by Spitters et al. [107]. More recently, Yao et al. [126] suggested new models for this separation.

11.2.2 Solar Tilted Radiation Models

Over a tilted surface, the global radiation is given by the sum of the three tilted components:

$$G_t = G_{Bt} + G_{Dt} + G_{Rt} \tag{11.4}$$

Each of the tilted components can be expressed as a simple function of the horizontal component:

$$G_t = R_b G_B + R_d G_D + R_r G_h \tag{11.5}$$

where R_b, R_d, and R_r are the direct, diffuse, and reflected factors, defined as the ratios between the respective horizontal and tilted radiation. Various solar transposition models have been developed in the literature to compute these three ratios, in order to calculate G_t.

The direct (or beam) radiation factor R_b is analytically computable given the geometric properties of the direct radiation. It must, however, be treated differently depending on the temporal resolution considered for the radiation. In case of an hourly computation, R_b is given by:

$$R_b^{hourly} = \max\left(0, \frac{\cos(\theta)}{\cos(\theta_Z)}\right) \tag{11.6}$$

with

$$\cos(\theta) = \sin(\beta)\sin(\theta_Z)\cos(\gamma_S - \gamma) + \cos(\beta)\cos(\theta_Z) \tag{11.7}$$

where θ and θ_z are, respectively, the angle of incidence on the tilted plane and the sun zenith angle, γ_s and γ are the sun azimuth angle and the aspect of the tilted plane (azimuth of the perpendicular direction to the plan), and β is the tilt angle of the plane. Note that if θ_z is not directly available, it can be computed using the following equation:

$$\cos\left(\theta_Z\right) = \sin\left(\phi\right)\sin\left(\delta\right) + \cos\left(\phi\right)\cos\left(\delta\right)\cos\left(\omega\right) \tag{11.8}$$

where ϕ, δ, and ω are, respectively, the latitude, the declination of the location, and the solar hour angle, expressing the time of the day [59]. In case of daily computation, R_b is given by Klein [67] and corrected by Andersen [5]:

$$R_b^{daily} = \frac{\int_{\omega_{sr}}^{\omega_{ss}} \cos\theta(\omega)d\omega}{\int_{\omega_r}^{\omega_s} \cos\theta_Z(\omega)d\omega} = \frac{R_b^1}{R_b^2} \tag{11.9}$$

where ω_r and ω_s are the sunrise and sunset hour angle, and ω_{sr} and ω_{ss} are the sunrise and sunset hour angles on the tilted surface. Detailed expressions for R_b^1 and R_b^2 are given in [67] and [5]. The above daily equation is also suitable for monthly mean daily estimations, which are widely used in potential studies.

The diffuse radiation factor R_d suffers from the stochastic behavior of the diffuse radiation (as it depends notably on cloud presence) and is computed empirically. Therefore, many diffuse models have been suggested in the literature. It is convenient to classify the models based on their temporal resolution, but also their isotropic or anisotropic assumption. We provide a list of suggested diffuse models in Table 11.1. Since comparisons of models have been presented in many different studies [33, 47, 89], our list is not exhaustive, but only a selection of what we consider to be the most used models.

The reflected radiation factor R_r is often computed using the hypothesis that the ground-reflected radiation is diffuse (meaning that the ground does not act like a mirror but rather reflects the incoming radiation in the form of multiple beams following a sphere), which results in the following isotropic model, given by Duffie and Beckman [38]:

$$R_r = \rho\left(\frac{1 - \cos\beta}{2}\right) \tag{11.10}$$

where ρ is the ground reflectance (or *albedo*) and β is the tilt angle of the surface. This expression was only rarely challenged by anisotropic reflection models [46, 111] that ultimately were not validated enough to be fully accepted in the domain. Therefore, the isotropic expression used in Eq. 11.10 is widely accepted to be a reasonable estimation of R_r.

Table 11.1 Diffuse tilted radiation models. H, D, and H&D mean that the model is suitable, respectively, for hourly, daily, and both hourly and daily estimations

Model	Year	Type	Time res.	R_d
Liu and Jordan [75]	1961	Isotropic	H&D	$\frac{1+\cos\beta}{2}$
Koronakis [69]	1986	Isotropic	H&D	$\frac{2+\cos\beta}{3}$
Tian et al. [113]	2001	Isotropic	H&D	$1-\frac{\beta}{180}$
Badescu [10]	2002	Isotropic	H&D	$\frac{3+\cos 2\beta}{4}$
Bugler [22]	1977	Anisotropic	H	$\frac{1+\cos\beta}{2}+0.05\frac{G_{Bt}}{G_D}\left(\cos\theta-\frac{1}{\cos\theta_z}\left(\frac{1+\cos\beta}{2}\right)\right)$
Temps–Coulson [111]	1977	Anisotropic	H	$\frac{1+\cos\beta}{2}\left[1+\sin^3\left(\frac{\beta}{2}\right)\right]\left[1+\cos^2\theta\sin^3\theta_z\right]$
Klucher [68]	1979	Anisotropic	H	$\frac{1+\cos\beta}{2}\left[1+F'\sin^3\left(\frac{\beta}{2}\right)\right]\left[1+F'\cos^2\theta\sin^3\theta_z\right]$ with $F'=1-\left(\frac{G_D}{G_h}\right)$
Perez [93]	1987	Anisotropic	H	$F_1\frac{a}{b}+(1-F_1)\frac{1+\cos(\beta)}{2}+F_2\sin(\beta)$ with F_1, F_2, a, b defined in [93]
Wilmott [121]	1961	Anisotropic	H&D	$R_b\frac{G_{Bn}}{1367}+C_\varphi\frac{1367-G_{Bn}}{1367}$ with $C_\varphi=1.0115-0.20293\beta-0.080823\beta^2$
Hay–Davies [51]	1961	Anisotropic	H&D	$AR_b+(1-A)\left(\frac{1+\cos\beta}{2}\right)$ with $A=\frac{G_B}{G_D}=\frac{G_h-G_D}{G_{oh}}$
Skartveit-Olseth [104]	1961	Anisotropic	H&D	$AR_b+\Omega\cos\beta+(1-A-\Omega)\left(\frac{1+\cos\beta}{2}\right)$ with $A=\frac{G_b}{G_{oh}}=\frac{G_h-G_D}{G_{oh}}$ and $\Omega=\max(0,[0.3-2A])$
Reindl [100]	1961	Anisotropic	H&D	$AR_b+(1-A)\left[\frac{1+\cos\beta}{2}\right]\left[1+\sqrt{\frac{G_B}{G_h}}\sin^3\left(\frac{\beta}{2}\right)\right]$ with $A=\frac{G_b}{G_{oh}}=\frac{G_h-G_D}{G_{oh}}$

11.2.3 Application

Concerning global horizontal radiation, numerous empirical models were used to fill the gap from the lack of measured values. A variety of linear models were used in the literature. In [1], the Angstrom model is used but with the parameters a and b (in Eq. 11.2) seasonally changed, which adds more accuracy to the model. In [102], data from Bangkok, Thailand, was used to develop a linear model for both global and diffuse solar radiation. Yohanna et al. [127] use the Angstrom model along with another model called the Hargreaves model to develop two global solar radiation models for three different locations in Nigeria. Clearness index was introduced as a function of the possible sunshine hours in order to calculate the model parameters. A more general linear model for global solar radiation for Malaysia based on Angstrom is developed in [66] and compared to various nonlinear and neural networks models. A model for the city of Istanbul in Turkey is proposed in [114], using sunshine ratios to capture global solar radiation. In [39], ambient temperature, relative humidity, and sunshine hours along with a nonlinear form of the Angstrom models are used to model global horizontal solar radiation in the city of Jeddah in Saudi Arabia.

Concerning tilted radiation, the direct and reflected components were computed in many locations using the two models presented in Eqs. 11.6 or 11.9, and 11.10, respectively, for the direct and reflected tilted radiations [39, 63]. These models are staple models that are almost always used. The diffuse tilted radiation, however, can be estimated by a large amount of models and represent the biggest challenge in the estimation of the tilted radiation. As a result, a few studies have compared the performance of multiple models in different settings. Noorian et al. [89] compare 12 different models for hourly diffuse radiation over a tilted surface, in the city of Karaj, Iran. The models include Badescu, Tian, Perez, Reindl, Koronakis, Skartveit and Olseth, Hay, Klucher, Temps and Coulson, and Liu and Jordan (see Table 11.1), and two other models not presented in the present review. The relative root-mean-square error is computed separately for the different directions of the considered tilted surfaces. For south-facing surfaces, the Skartveit and Olseth model outperformed the other models, with an RMSE of 10.16%. The Temps and Coulson model achieved the worst performance, with an RMSE or 54.89%. For west-facing surfaces, the performances of all models were generally lower, and the best accuracy is achieved by the Perez model, with an RMSE of 30.71%. Diez et al. [35] propose a similar comparison of models to estimate the diffuse tilted radiation over a south-facing surface in the city of Valladolid, Spain. Measurement data for both daily and hourly values is used to compute the resulting errors from the models. Out of ten models, the Reindl model achieved the second best performance, for both daily and hourly estimation, after the Muneer model [85]. Seven diffuse models are tested in [78] with solar radiation data measured over tilted surfaces on the EMPA campus in the city of Dübendorf, Switzerland. The data was recorded hourly from two 25 days periods in October and March/April and is claimed to capture various atmospheric conditions and sun altitudes. The best performance was achieved by the Perez model. It finally appears that depending on the location of interest, and the level of precision of the input data, the performance of the different models is variable. Some models, however, including the Reindl model (for daily estimation) and the Klucher and Perez models (for hourly estimation) give on average good results.

11.3 Geostatistical Methods

11.3.1 Principle

Geostatistical methods, or *geostatistics*, are a powerful and well-established family of methods to perform statistical analysis on spatial data. In particular, geostatistical methods have been extensively used with environmental data, including wind or solar data. In the case of PV potential estimation, geostatistics can be used for the spatial interpolation of solar horizontal radiation, leading to solar radiation maps, as shown in Figs. 11.2, 11.3, and 11.4.

In geostatistics, spatial points are considered as single realizations of a random variable of interest Z, for instance solar radiation. In a general setting, we consider that Z has a deterministic part and a residual stochastic (probabilistic) part, as follows:

$$Z(\mathbf{x}) = m(\mathbf{x}) + S(\mathbf{x}) \tag{11.11}$$

where $\mathbf{x} = (x, y)$ is a two-dimensional location vector, $m(\mathbf{x})$ is the deterministic part, the mean of $Z(\mathbf{x})$ (to be estimated preferably with an adapted physical model), and $S(\mathbf{x})$ is the stochastic part. Geostatistical prediction is traditionally performed by a classical family of linear models called *kriging* models [119].

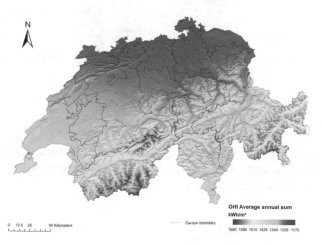

Fig. 11.2 Global Horizontal Irradiance (GHI) yearly map, in Switzerland. This figure is reused via Elsevier License #4134701271907, from [9]

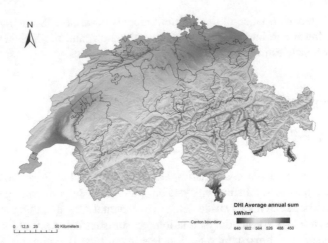

Fig. 11.3 Diffuse Horizontal Irradiance (DHI) yearly map, in Switzerland. This figure is reused via Elsevier License #4134701271907, from [9]

Fig. 11.4 Extraterrestrial Horizontal Irradiance yearly map, in Switzerland. This figure is reused via Elsevier License #4134701271907, from [9]

The principle of kriging is as follows: Assuming the probabilistic behavior of the random field, one can extract the possible realizations of the field (meaning the values taken by the field) that agree with the data. The interpolation is performed using a local probabilistic weighted average of known neighborhood points, to make a spatial prediction for a new point $\mathbf{x_0}$ [64]:

$$\hat{Z}(\mathbf{x_0}) = \sum_{i=1}^{N} w_i(\mathbf{x_0})Z(\mathbf{x_i}) + w_0(\mathbf{x_0}) \tag{11.12}$$

where the w_i are the kriging weights, N is the number of measurements, and x_i are the N measured neighborhood points. If N (number of measurements) is too large, a local neighborhood can be used, chosen with appropriate knowledge or with a k-Nearest Neighbors (k-NN) model. The kriging weights are originally unknown and are the main output of the method, summarized in the following part.

As for any probabilistic method, a crucial point is to choose a model for Z. The choice of the model is based on the spatial correlation between points, captured by the so-called *semi-variogram* [64]:

$$\gamma(\mathbf{h}) = \frac{1}{2} \text{var} \left[Z(\mathbf{x}) - Z(\mathbf{x} + \mathbf{h}) \right] = \frac{1}{2} \mathbb{E} \left[(Z(\mathbf{x}) - Z(\mathbf{x} + \mathbf{h}))^2 \right] \tag{11.13}$$

where var and $\mathbb{E}$ are the variance and expectation operator, respectively. $\gamma(\mathbf{h})$ should be, in general, a function of $\mathbf{x}$ as well; however, the hypothesis of second-order stationarity, considered in geostatistics, implies that $\gamma(\mathbf{x}, \mathbf{h}) = \gamma(\mathbf{h})$. The empirical estimate of the semi-variogram is used in practice [64]:

$$\gamma(\mathbf{h}) = \frac{1}{2N(\mathbf{h})} \sum_{i=1}^{N(\mathbf{h})} \left(Z(\mathbf{x_i}) - Z(\mathbf{x_i} + \mathbf{h}) \right)^2 \tag{11.14}$$

where $N(\mathbf{h})$ is the number of pairs of measurement points separated by vector $\mathbf{h}$. Note that in the considered model for Z defined in Eq. 11.12, the mean of the Z is equal to its deterministic part, and thus the mean of the residual is zero: $\mathbb{E}[Z(\mathbf{x})] = m(\mathbf{x})$, $\mathbb{E}[S(\mathbf{x})] = 0$. The linear estimate $\hat{Z}(\mathbf{x})$ is desired to be (i) optimal, meaning as close to $Z(\mathbf{x})$ as possible, and (ii) unbiased, meaning that the estimated and the measured values are, on average (meaning under the expectation operator), equal. Condition (ii) implies the following:

$$\mathbb{E} \left[\hat{S}(\mathbf{x}) \right] = \mathbb{E} \left[\hat{Z}(\mathbf{x}) - Z(\mathbf{x}) \right] = 0 \tag{11.15}$$

where $\hat{S}(\mathbf{x})$ is the estimate of the residual $S(\mathbf{x})$ and condition (i) is achieved when the variance of the error is minimized [64]:

$$\min \left\{ \sigma_R^2(\mathbf{x}) \right\} \tag{11.16}$$

where:

$$\sigma_R^2(\mathbf{x}) = \text{var} \left[\hat{S}(\mathbf{x}) \right] = \mathbb{E} \left[\left\{ \hat{S}(\mathbf{x}) - \mathbb{E} \left[\hat{S}(\mathbf{x}) \right] \right\}^2 \right] \tag{11.17}$$

The previous Eqs. 11.15 and 11.16 form a system of equations that will provide the kriging weights for any type of kriging model. The practical steps to perform kriging are as follows:

- Compute the empirical semi-variogram from Eq. 11.14 with the observed (measured) data.
- Fit a model with the obtained semi-variogram (try typical models like spherical, Gaussian, exponential).
- Use the previously fitted model to krige: (i) Compute the covariance function of the model for measured and unknown data, (ii) obtain the kriging weights by solving Eqs. 11.15 and 11.16, and predict the value of the field at a new point x_0 with Eq. 11.12.

Various types of kriging have been developed depending on different hypotheses of the model. These include: *simple kriging* (SK—the mean is constant and known), *ordinary kriging* (OK—the mean is constant but unknown), *universal kriging* (UK—the mean is a low-degree polynomial function), *kriging with external drift* (KED—to allow the use of a secondary variable in the estimation, for instance elevation). Another interesting model that accounts for external explanatory variables is called *residual kriging* (RK). In this type of kriging, the previously detailed steps are not used on $Z(x)$ directly, but instead, a regression analysis between $Z(x)$ and the chosen external variables $a_i(x)$ is first performed:

$$Z(x) = Z_r(x) + \varepsilon(x) = \sum \alpha_i a_i(x) + \varepsilon(x) \qquad (11.18)$$

where $\varepsilon(x)$ is stochastic noise. We then perform kriging on the residual variable $r(x) = Z(x) - Z_r(x)$ that does not capture the variability caused by the external variables $a_i(x)$, but focus on the residual variability. The resulting estimation $\hat{r}(x)$ is then added to the regression model to obtain the final estimate:

$$\hat{Z}(x) = Z_r(x) + \hat{r}(x) \qquad (11.19)$$

A more detailed presentation of geostatistics in general and kriging models can be found in [64].

11.3.2 Application

Geostatistics has been used extensively in the literature to allow for spatial solar radiation interpolation. These methods include mainly kriging techniques, among which ordinary kriging, universal kriging, and residual kriging were the most used. D'Agostino and Zelenka [31] used kriging with observed data from the Swiss meteorological network to predict global horizontal radiation mapping in Switzerland. The estimates are validated with satellite images from Meteosat and achieve an RMSE of 2.5 MJ m^{-2}. In [99], ordinary kriging is used to plot contour maps for global solar radiation in Saudi Arabia. They use 41 measurement stations to build and cross-validate the model. The mean percent errors for the estimates were around 1%. Universal kriging was used in [40] for mapping global solar radiation with RMSE values for the estimates ranging from 1.4 to 10 MJ m^{-2} day^{-1}. A comparative analysis of ordinary and residual kriging methods for mapping global solar radiation in southern Spain (Andalusia) is presented [3]. Measurement data collected at 166 stations (during 4 years, from 2003 to 2006) is used. The results showed a certain superiority of residual kriging over ordinary kriging, the first accounting for topography, while the latter does not. Another comparative study, this time between ordinary kriging, ANN models (reviewed in Sect. 11.7.2), and other models, was proposed in [84] to estimate the daily global solar irradiation over Spain. Ultimately, the ANN models showed the best performance. More recently, Inoue et al. [58] suggested the use of cloud movement information as an extra variable to map solar radiation with kriging.

11.4 Constant-Value Methods

11.4.1 Principle

Constant-value methods consist of using a scalar multiplier that can be applied to the entire region of interest in order to account for one or multiple constraints in the estimation of the solar rooftop PV potential. Often, this constant multiplier is simply assumed based on rules of thumbs. It is often used to estimate the percentage of rooftops with desirable conditions for PV installations. These conditions include suitable slope and aspect configurations, low shadowing effects, and large available rooftop area for PV installation (with no or only a few obstructing elements such as chimneys or HVAC systems). In some cases, the constant multipliers are chosen not solely by assumption but using available data on the population or the building characteristics of the region. Popular strategies consist in using the population density of the area or building types proportions (residential, commercial, industrial, etc.) to compute the multipliers.

11.4.2 Application

Constant-value methods are very easy to use and not computationally expensive. As a result, they have been quite popular over the last two decades. They are also suitable for a very large-scale potential analysis (e.g., large country or a whole continent) and can also be used to provide a first approximation of the potential.

Constant-value method has been widely applied to directly provide an estimation of the available rooftop area for PV installation. Several studies assume one coefficient to account for the available rooftop area and apply it to an entire region. In [117], building vector data from the Israeli Central Bureau of Statistics is used to estimate the rooftop area suitable for PV installation for different types of buildings (residential, commercial, etc.). After an estimation of the total available area for each building type based on GIS analysis, the authors consider a first scenario in which they use a constant value of 30% to account for the rooftop available space for all types of buildings. This results in a yearly potential of 15.9 TWh, which represents 32% of the Israeli electricity demand. A second scenario is considered, where only rooftops with a rooftop area higher than 800 m^2 are taken into account. In this case, a value of 50% is applied to these rooftops to account for the available rooftop space, which corresponds to only 7% of the electricity demand in Israel. Other studies assume multiple coefficients to account for various constraints (including solar accessibility, shading effects) and combine them to derive the final potential. IEA [48] has suggested a method to estimate the rooftop PV potential in various countries in Europe. Depending on the country, constant coefficients are assumed to account for architectural suitability (including obstructing superstructures, shading impacts, historical considerations), solar suitability (amount of solar radiation available over the rooftop throughout the year), along with a typical PV efficiency and other factors to directly estimate the potential. For instance, it is estimated that Japan can only cover 14.5% of its electricity demand, while Switzerland and Spain can produce the equivalent of 34.6% and 48% of their annual electricity demand with rooftop PV technologies.

It has been shown that in most countries, the general rooftop area is significantly depending on the roof, flat or pitched. In [25], the proportion of rooftop available area in the USA is estimated to be 18% and 65%, respectively, for pitched and flat roofs. Therefore, many studies assume a constant coefficient for each of the two cases (flat and pitched buildings) in order to estimate the available rooftop area for PV installation. While some studies assume the proportion of flat and pitched buildings, most estimate the number of flat and pitched buildings and simply apply two different coefficients accordingly to the computed proportion. In a recent study, Rodriguez et al. [101] use 3D city models to differentiate between flat and pitched rooftops, and use different coefficients accounting for various constraints (shadow effects, orientation losses, construction restrictions, separation of panels), with a pitched version and a flat version for most of them. A construction restriction factor (accounting for chimneys, HVAC systems, etc., already installed over the rooftop) is, for instance, chosen to be 0.8 for flat rooftops and 0.9 for pitched rooftops.

Many studies attempted to estimate the available rooftop area using the known footprint area (ground floor area). Such studies include National Renewable Energy Laboratory (NREL) reports [34, 43, 91] that base their estimation on floor-space data from McGraw-Hill. Assumptions are made about the proportion of tilted and flat rooftops, depending on the type of the building. As it is estimated that a minority of residential buildings is flat (8%), and a majority of commercial buildings is flat (63%), shading impacts alone are accounted for commercial buildings, while slope, aspect, and shading are accounted for residential buildings. An average of 62.5% and 24.5% of rooftop area is estimated to be available, respectively, for commercial and residential rooftops. Other similar studies estimating the available area by applying coefficients to the total footprint area include [32, 48, 76, 92]. Most of these studies use different coefficients based on different building types.

Another strategy to estimate the various rooftop constraints is to use population data and estimate the available roof area based on identified relationships between the two variables. Lehmann and Peter [72] stated that there was a relationship between the available roof area and the population density in the European Union. The municipalities were separated in three population density categories: low, medium, and high, respectively, corresponding to a density of under 100 inhabitants/km^2, 100–500 inhabitants/km^2, and above 500 inhabitants/km^2. The authors found a linear correlation with an R^2 coefficient of 0.993 stating that available roof area $= 70 \cdot$ population $+ 237000$. This linear function was found based on their specific data covering the E.U. Even though the correlation coefficient R^2 is very promising, this relation does not necessarily reflect the local patterns of E.U regions or countries considered independently. It was confirmed by Schallenberg-Rodriguez [103], showing that the relationship was not true for the Canary Islands, after comparing with Spanish cadastre data. It does not, however, reject the possibility of finding a local relationship in the region of interest, based on local data. Wiginton et al. [120] extracted the same relationship and used it as an additional information in their GIS-based potential study (see Sect. 11.6.2). In all cases, there was a confirmed positive correlation between the two variables. Other studies used the population density not as a direct predictor but rather as a classification factor, as it is the case in [60] (further reviewed in Sect. 11.5.2) to classify municipalities.

Besides rooftop available area itself, numerous studies have applied all sorts of coefficients to separately account for shading effects and other constraining factors [60, 72, 94, 120, 128]. Nevertheless, they show great variation, depending on the location of interest, even within one region or one city. For example, a study in Switzerland carried by Montavon et al. [83] found three significantly different utilization ratios for three different neighborhoods (0.49, 0.73, 0.95). As a result, the use of such coefficients needs to be considered with caution for extrapolation, and the coefficients should be validated with local data, for instance using a sampling strategy, presented in the following section.

11.5 Sampling Methods

11.5.1 Principle

The idea of sampling methods is to compute a variable of interest solely for a chosen sample of points, or locations, and use an adequate strategy to extrapolate it to the whole data set, or whole region. It allows to provide a more reliable estimate of the variable than an assumed constant coefficient, while keeping the computational requirements of the method reasonably low. Note that the use of a constant coefficient is still common with sampling methods, yet it is no longer assumed and should be based on a thorough analysis of samples. While this method is in theory usable for any kind of variable, it has been mainly used in large regional settings to estimate the rooftop available area. The reasons are multiple: (i) It is relatively easy to obtain local data to estimate the available area in small areas, for example with aerial images, or precise building shape files, (ii) the extrapolation strategy can be based on simple rules, such as building types or based on the known values for footprint area or population density, and (iii) while the available area estimation is often considered to be the most important factor in a potential study, a precise computation of the total available area can be very computationally expensive for a large region. The area can be either estimated purely geometrically first, independently of slope, aspect, and shading considerations, either together with these considerations, in order to directly estimate the complete available area over the rooftops. We will present in the literature review some of most significant studies that used such a sampling method.

11.5.2 Application

One of the first studies, to the best of our knowledge, that used a sampling method to estimate the total roof surface in a large-scale estimation is Izquierdo et al. [60], in Spain. The study classifies the Spanish municipalities based on their building and population density that they both express from low to very high. They define a density class (or "representative building typology") for each possible building and population density couple. They compute a different total available area ratio (that includes utilization architectural suitability, shading effects, and some additional suitability factor) for each representative building typology, based on samples of building for each typology. In each sample, GIS vector data is used to compute the ratios, and various steps are applied to verify the quality of the sample, including the computation of the coefficient of variation of the sample. An error term is also computed to measure the uncertainty of the sampling strategy for each typology. They further extrapolate the ratios for all the municipalities in Spain. It results in a total available area of $571 \pm 183 \, \text{km}^2$ in the whole Spain. Schallenberg-Rodriguez [103] uses a similar strategy to compute the rooftop PV potential in the Canary Islands, using the total roof area and various factors to account for the availability based on different building and roof typologies. The results are compared with Izquierdo et al. [60] results obtained for the Canary Islands, and while they are not identical due to different assumptions, they generally agree. Ordonez et al. [90], in a rooftop PV potential study in Andalusia, Spain, use aerial images from Google Earth and construction data to compute the available area for a sample of buildings. They use a similar sampling strategy than in [60]. The sampled building footprints are digitized in 3D models in AutoCAD from Google Earth, and superstructures obstructing the rooftop area (HVAC systems, chimneys, etc.) are manually erased from rooftop. The results show that around 53% and 18% of the roof area are available, respectively, for flat and pitched roofs, in Andalusia. More recently, Bocca et al. [16] presented a sampling strategy to estimate the PV potential of a large area where only a few and scattered measurement points are available. They achieved a mean absolute error of less than 6%.

Various other studies used sampling in order to extrapolate availability factors to large regions. However, the quality of samples is not always thoroughly verified, and the extrapolation strategy is often very simple. In [6], the PVWatts Calculator, allowing users to draw a PV system over a rooftop and calculating the PV potential, is used in order to compute the potential for a sample of buildings in New York City. The obtained available area percentage is then used for various locations in the city in order to compute their potential for rooftop PV solar energy. In another study to compute the rooftop PV potential in Kingston, Nguyen and Pearce [87] use aerial images to extract the available roof area over a sample of buildings. An additional percentage of available area is removed in case of pitched roofs. They conclude that about 33% of the rooftop area is suitable for PV installation. Eventually, sampling methods can be a very useful tool to extract local rooftop characteristics in a region of interest, without the requirement of a very large data set and high computational power. Nevertheless, the quality of the samples must be verified statistically, and the extrapolation strategy needs to be validated. An estimation of the resulting uncertainty is also desirable.

11.6 GIS and LiDAR Data-Based Methods

11.6.1 Principle

Geographical Information Systems (GIS) are a set of geographical data and tools allowing to perform spatial analysis. ArcGIS and QGIS are among the most popular GIS software. GIS can be used to compute some variables of interest (e.g., rooftop area, footprint area, roof azimuth, roof slope) as well as solar PV potential. Building geometric characteristics including building rooftop is usually derived from orthophotagraphies or light detection and ranging (LiDAR) data. LiDAR is a very popular remote sensing technique that allows for the sampling of the earth accessible surface in a form of a grid of elevation points. It uses laser light (generated by drones or small planes) in order to give, at a chosen resolution, elevation values of the earth surface, naturally accounting for all kinds of obstacles, including buildings, trees. As a result, LiDAR data is very suitable to model building rooftops, provided that the resolution is high enough. GIS can be combined with LiDAR to build powerful methods for rooftop PV potential estimation.

GIS-based methods are very separated from coefficient-based or sampling methods by their basic estimation strategy. The two latter methods assume extrapolating ratios or consider samples of examples to derive more informative coefficients, again to further extrapolate. These methods were naturally driven by the general lack of data and computational power. GIS methods, in contrast, attempt to directly extract the suitable PV surfaces from all buildings in the considered region. Consequently, they are more computationally expensive, but theoretically considerably more accurate. This is the key of their recent popularity, as they take advantage of the increasing availability to high computational power. Besides, LiDAR data has become more widely available in high resolutions over the last few years, allowing for a more precise estimation of the rooftop characteristics at various scales.

GIS-based methods can be used to compute the various suitability variables needed for a PV potential study, in an automatic fashion. These variables include primarily available roof area, shading effects, but also slope and aspect values of rooftops, rooftop shapes, and tilted solar radiation over rooftops. GIS methods can be also used to directly derive the most suitable areas for PV, without a precise estimation of each suitability variable. This is achieved by automatically discarding roofs with undesirable characteristics, including unsuitable slope and aspect values, highly shaded rooftops, or with low incoming radiation.

11.6.2 Application

There have been numerous GIS-based methodologies for PV potential estimation using high-resolution images and LiDAR data in recent years. We present here some of the most significant ones. We will first review some complete PV potential studies and will then review a few studies focusing only on variables independently. These include shading effects and slope, aspect, and roof shapes analysis, particularly adapted for LiDAR data.

Several studies estimated the rooftop PV potential at a neighborhood scale. In these studies, LiDAR data is used to determine slope and aspect of all rooftops individually [65, 88]. Tereci et al. [112] used LiDAR data together with a modeling software to build a Digital Surface Model, in order to classify roof types and filter rooftops suitable for PV based on typical optimal slope and direction conditions. The annual solar PV potential is obtained by using average annual global radiation values. Jochem et al. [61] develop an automatic detection of rooftops suitable for PV using only LiDAR point data for a set of 1071 rooftops. They use an annual radiation threshold of 700 kWh/m² to characterize suitable rooftops for PV installations. Hofierka and Kanuk [53] used a 3D city model created in GIS together with the PVGIS radiation tool from [110] to estimate the PV potential for the building rooftops of a small city. Bill et al. [15] developed a 3D model for solar potential estimation on building envelopes based on LiDAR data. Lukac et al. [79] estimate the average daily production of electrical energy throughout the year, using LiDAR data to account for the precise shape of the rooftops. Efficiency characteristics for different types of photovoltaic modules are used to estimate the final technical potential for PV. Brito et al. [21] used LiDAR data to compute the rooftop PV potential of 538 buildings, using filtering to detect roofs with no shading, and desirable slope and aspect conditions. The solar radiation was computed with the Solar Analyst tool from ArcGIS. Redweik et al. [98] develop a Digital Surface Model for the campus of the University of Lisbon, to assess the solar potential of roofs and facades, at a spatial and temporal resolution of, respectively, 1 m and 1h. Besides the roof shape, an algorithm was developed to account for shading effects over the roofs and facades.

A number of potential studies using GIS strategies can also be found at a relatively larger scale. Bergamasco and Asinari [13] estimated the available area for PV in each municipality of the Piedmont Region in Italy, using local vector polygon data with building typologies. They also add a few coefficients to account for missing information in the GIS data, like the shadowing coefficient, that they take from [60]. They use a solar database from the Joint Research Center of the European Commission from 2010 to extract radiation values and finally provide a potential estimation. The study was further improved in [14] by using extra orthophotos and applied to the Turin region. Ultimately, it is concluded in the paper that the available area results show a total error of 1.7%. Vardimon [117] use orthophotos together with GIS in order to compute the total rooftop area for PV in Israel. The study is further extended to estimate the total available rooftop area by the use of coefficients. Building polygons were extracted for each of the 1,200,000 buildings in the entire country simply by analyzing the orthophotos with photogrammetric methods. Gagnon et al. [44], in an NREL study, use LiDAR data together with GIS to estimate the rooftop potential areas for PV in 128 cities in the USA. The suitable rooftop areas are computed by applying suitability criteria depending on shading effects, slope and azimuth values (detecting the rooftops with desirable characteristics). These variables are computed based on LiDAR data. Furthermore, statistical models are developed to extend the potential values to the entire US continent.

Some recent studies have been developing models using LiDAR data to extract the roof shapes of buildings, for solar potential estimation (often when the resolution of the data is not high enough to extract the exact slope and aspect values of rooftops). Boz et al. [18] use building footprints GIS data together with LiDAR data to extract the roof shapes of buildings in Philadelphia, USA. The rooftops are classified based on roof characteristics (seven slope classes, five aspect classes, and six different building types), and rooftop polygons are created accordingly to the classes over the footprints. Monthly shading values are computed. The total ground floor area suitable for PV systems is finally identified (33.7%). Gooding et al. [45] use low-resolution LiDAR data to extract building roof shapes by classifying them among a catalog of common roof shapes. The best fit roof shape is selected according to various geometrical characteristics. The classification is performed for 705 building in Leeds, UK, and validated with a ground-based survey for 169 of them and aerial photographs for the 536 remaining ones. The authors find a final classification accuracy of 87%.

Several studies use LiDAR data to analyze shading impacts on the building rooftops, which remain one of the biggest challenges to provide a precise potential study. These shading studies, however, were still applied for relatively small-scale case studies, due to the high computational requirements. Levinson et al. [73] use high-resolution LiDAR data to compute a Digital Surface Model including all trees and buildings in residential areas. The hourly shadowing over the rooftops is computed. The roofs are divided into parcels, and raster hillshade values are computed (with an ArcGIS toolbox) over the rooftops to compute the shading impact of neighbor obstacles and other rooftop parcels. Tree growth is also simulated to provide future shading values. The fraction of radiation lost to shading is finally computed over the rooftops. Nguyen and Pearce [86] develop an algorithm that uses LiDAR data to compute the shading effects over the rooftops of 100 buildings in Kingston, Ontario. They compute the shading impact over the direct radiation with hillshade calculations, including both shadowing from terrain and neighbor obstacles. The shading impact over the diffuse radiation is computed by an estimation of the Sky View Factor (SVF). The irradiation is modeled by the r.sun software. Note that the methodology is based on the work by Carneiro et al. [24] and Ratti et al. [97]. More recently, Hong et al. [54] suggested another study focused on a shading analysis to compute the rooftop available areas in a district of Seoul, South Korea. Hillshade analysis was performed on the 15th of each month, for each hour. In order to compute the available area, shaded portions of the roofs (characterized by a hillshade of 0) were discarded, and roof portions smaller than 33 m^2 were discarded as well (as they are considered too small to accommodate PV installations). The technical PV potential was finally computed combining hourly radiation from the World Radiation Data Centre with the available area values. Assouline et al. [9] use LiDAR data to extract shadowing variables over rooftops in the canton of Geneva, Switzerland. Geneva canton is one of the 26 different cantons in Switzerland, which define the biggest administrative divisions in the country. It includes 46 communes, including the city of Geneva. A methodology combining GIS and hillshade analysis over LiDAR data was proposed in order to compute the monthly shading impact from terrain and neighboring buildings and obstacles. The LiDAR data was available in the form of a Digital Orthophoto Map (DOM) with a resolution of 2×2 [m^2]. In particular, the shaded portions of rooftops were identified using the Raster Calculator within the Spatial Analyst toolbox in ArcGIS (hillshade = 0), and the average hillshade of the remaining roof areas was computed. An illustration of the process is shown in Fig. 11.5.

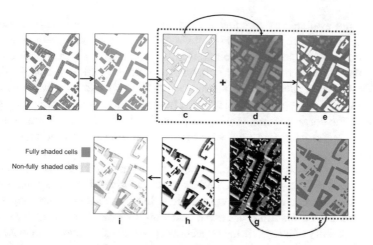

Fig. 11.5 Schematic presentation of a methodology to compute shading factors. **a** Buildings with detailed roof geometry, **b** dissolve detailed roof geometry into a continuous polygon, keeping the same outer boundaries as the original polygons, **c** reverse vector map of buildings as void and the surroundings as filled polygons, **d** Digital Orthophoto Map (DOM) with a 2 m by 2 m resolution, **e** "negative" DOM raster map extracted by clipping the DOM (**d**) over the "reverse" building polygons (**c**), **f** a Boolean raster map or IsNull raster map, Null cells (void cells) assign to value 1 indicating there are buildings, whereas not Null cells assign to value 0 indicating there are no buildings, **g** hillshade map for buildings and their landscape surroundings in urban areas, **h** clipped hillshade map for buildings extracted by clip hillshade map (**g**) over IsNull raster map (**f**), **i** binary raster map showing cells that are nonfully shaded in yellow and cells that are fully shaded in blue. This figure is reused via Elsevier License #4134701271907, from [9]

11.7 Machine Learning Methods

11.7.1 The Basics of Machine Learning (ML)

Machine learning (ML) methods are based on algorithms that improve their performance with increasing experience. Experience, in this case, is primarily the information provided by examples. For a more formal formulation, let us define input variables $X_1, X_2, \ldots, X_d$ of interest for the studied phenomenon. These variables might be temperature, precipitation, etc. Each data point is one realization of these variables, forming a p-dimensional vector of values taken by the input variables $(x_1, x_2, \ldots, x_d) = \mathbf{x}$. As a result, an input dataset of N points can be seen as an $N \times p$ matrix formed by N rows $(\mathbf{x}_1, \mathbf{x}_2, \ldots, \mathbf{x}_N)^\top$. Similarly, let us define an output variable (e.g., solar radiation) Y that takes a value y for each data point, thus forming the values $(y_1, y_2, \ldots, y_N)$. Note that the input variables are sometimes called *features* or *predictors*, data points are *samples* or *instances*, the output variable is the *target*, and the target values are *labels*. In a classical supervised learning task, we know both feature and target values for a certain amount of data points. There are, however, other settings, namely unsupervised and semi-supervised learning (where labels are not available or very few are available) which will not be discussed here. This gathering of observed points is known as the *labeled set*, or sometimes simply the *learning set*, and consists of couples $(\mathbf{x}_1, y_1), \ldots, (\mathbf{x}_N, y_N)$. A machine learning task (in the supervised framework) aims at learning a function $\varphi : \mathcal{X} \to \mathcal{Y}$, where $\mathcal{X}$ and $\mathcal{Y}$ are, respectively, the input and output spaces (where the input and output variables live), based on the labeled set. The goal is to obtain predictions $\varphi(\mathbf{x})$ as close as to the corresponding target y as possible. If $\mathcal{Y} = \mathbb{R}$, it is a regression task, and φ is a *regressor*; if $\mathcal{Y}$ is a finite set of classes, it is a classification task, and φ is a *classifier*.

One of the main concerns in ML is of course to maximize the performance of a learned model φ. The performance of the model is given by a loss function L that measures the discrepancy between predicted output values and known target values. Many different loss functions are used, based on the task of interest. In a statistical sense, the quantity of interest is the expected prediction error, over all possible values of $\mathcal{X} \times \mathcal{Y}$. The fact of using all possible values in the product space $\mathcal{X} \times \mathcal{Y}$ is crucial: The finality is not to learn the model perfectly in the labeled set (known set), but to assure that this model is applicable outside of the labeled set, to allow the prediction of labels for new points. However, since we generally do not know the distribution of the input and output variables, we generally approximate the expected prediction error with the *test sample estimate* of the error. The principle of the test sample estimate is simple: (i) Separate the labeled set into a *training set* and a *test set*, (ii) train a model only using the training set, (iii) predict the output values for points in the test set, and compare with the actual labels to compute the test error. A typically used proportion is 75% of the labeled set for the training set and 25% for the test set. Note that all models have parameters, usually called *hyperparameters*, that the user has to choose in order to maximize the performance for a given task. Therefore, a part of the training set can be used as a *validation set*, to try compared models trained with different parameters. However, this parameter tuning step is often achieved at the same time as the training itself, through a procedure called *K-fold cross-validation* [108], depicted in Fig. 11.6.

There exist multiple functions to compare known labels and predicted values for the test sample estimate of the error [122]. Although some of these functions are attached to certain problems as the "standard" measures of error, all can be used to capture the different discrepancies of the model. Some of the most used are presented here. In all the following definitions, y^{obs} is the observed (known) output, y^{pred} is the predicted output, and N_{test} is the size of the testing set, meaning the number of data points for testing, used to compute the model error.

The *Root-Mean-Square Error* (RMSE) is the standard error of estimation for regression. It is the square root of the mean square error:

$$\text{RMSE} = \sqrt{\frac{\sum_{i=1}^{N_{\text{test}}} \left(y_i^{\text{pred}} - y_i^{\text{obs}}\right)^2}{N_{\text{test}}}} \tag{11.20}$$

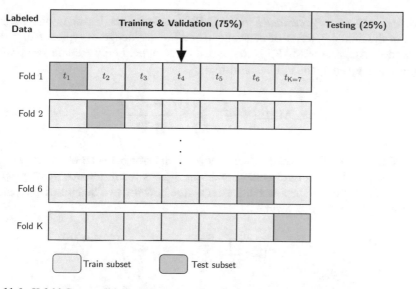

Fig. 11.6 K-fold Cross-validation illustration, here for K = 7. The training data is first separated in K equal parts. It is then going to be used K times (corresponding to K different folds of the data), in each of which one of the K parts is used to test the model trained with the K − 1 remaining parts. The RMSE is stored for each fold, and the mean RMSE is computed. This whole process is done for multiple sets of parameters. The "best" set of parameters is then the one offering the lowest mean RMSE

The RMSE expresses the average error between observed (i.e., known, measured, or validated in some way) and predicted output, in the order of magnitude of the quantities of interest. As a result, it only has sense if compared to typical values of the data.

The *Normalized Root-Mean-Square Error* (NRMSE) is used to neutralize the relativity of the RMSE. It is simply normalizing the RMSE by the average observed value:

$$NRMSE = \frac{RMSE}{\overline{y^{obs}}} \tag{11.21}$$

where $\overline{y^{obs}} = \left(\sum_{i=1}^{N_{test}} y^{obs} \right) / N_{test}$.

The *Mean Absolute Error* (MAE) measures the average discrepancy between known and predicted values, as RMSE does, but using the absolute value, instead of the root-mean-square:

$$MAE = \frac{1}{N_{test}} \sum_{i=1}^{N_{test}} \left| y_i^{obs} - y_i^{pred} \right| \tag{11.22}$$

The *Mean Absolute Percentage Error* (MAPE) is a normalized error, giving a general idea of the model in terms of percentage, and no knowledge of typical data values is needed. An issue of MAPE is that it can be computed only for nonzero positive values. It is given by the following formula:

$$\text{MAPE} = \frac{1}{N_{\text{test}}} \sum_{i=1}^{N_{\text{test}}} \left| \frac{y_i^{\text{obs}} - y_i^{\text{pred}}}{y_i^{\text{obs}}} \right| \tag{11.23}$$

The *Mean Biased Error* (MBE) is a simple mean error between observed and predicted output. The main difference is that the sign of the difference matters, which can be useful in some specific studies to capture how the two quantities compare. It is given by the following formula:

$$\text{MBE} = \frac{1}{N_{\text{test}}} \sum_{i=1}^{N_{\text{test}}} \left(y_i^{\text{pred}} - y_i^{\text{obs}} \right) \tag{11.24}$$

The *Accuracy Error* (AE) is used in a classification task, where the values are discrete, and there is no need for a measure between continuous values. The goal is to compute how many times the predicted label (class number) is the same as the actual label, as a percentage. As a result, the following indicator function is used:

$$1_{[x=y]} = \begin{cases} 1, & \text{if } x = y \\ 0, & \text{otherwise} \end{cases} \tag{11.25}$$

The Accuracy Error is then given by:

$$\text{AE} = \frac{1}{N_{\text{test}}} \sum_{i=1}^{N_{\text{test}}} 1_{\left[y_i^{\text{pred}} = y_i^{\text{obs}} \right]} \tag{11.26}$$

where y_i^{pred} and y_i^{obs} are in this case discrete classes.

11.7.2 Three Popular ML Algorithms

Three important machine learning methods will be presented here, representing three of the most common approaches in the field: Artificial Neural Networks (ANNs, referred nowadays as Deep Learning), Support Vector Machines (SVM, being part of the general family of kernel methods), and Random Forests (RF, being part of the general family of ensemble learning methods). These methods were also chosen as they are the most popular methods for energy applications and environmental data analysis in general, as presented further in Sect. 11.7.3. The three methods are presented here in a supervised learning setting.

11.7.2.1 Artificial Neural Networks

Artificial Neural Networks (ANNs) are one of the first approaches towards machine learning, as an attempt to build machines that learn from examples. Like any ML method, an ANN wants to build a function mapping the input features to the output values, by studying previously observed and stored data. Yet, the structure of ANNs is quite different from the other ML methods. ANNs were designed to mimic the learning process of humans. Consequently, the basic principle of ANN is to replicate the architecture of the human brain with interconnected series of learning neurons to learn data patterns. As a result, the resemblance with the brain is at two levels: The knowledge is acquired through a learning process, and this knowledge is stored in neurons that work as synaptic weights.

A schematic of a neuron in an ANN, following the classic McCulloch-Pitts model [80], is shown in the left side of Fig. 11.7. It is the building block of the ANN, as it acts as a filter between two consecutive nodes in the network. The general architecture of the network is shown in the right side of Fig. 11.7. The network is composed of hidden variables between feature inputs and the final output, gathered in hidden layers (one hidden layer is shown in the figure for simplicity, but it is possible to add many layers). The hidden layers express the complexity of the relation between the different variables and between inputs and the output variable. Between two layers (input-hidden, hidden-hidden, or hidden-output) are neurons that learn weights to express the local output v as a linear expression of the local inputs (u_1, u_2, u_3, if we follow the notations in the neuron in Fig. 11.7). The nonlinearity is expressed by applying a nonlinear function, so-called activation function, to the linear combination of inputs values, with an added constant. If we consider M inputs for a neuron, the local output v is expressed as:

$$v = f\left(\sum_{i=1}^{M} w_i u_i + b\right) \tag{11.27}$$

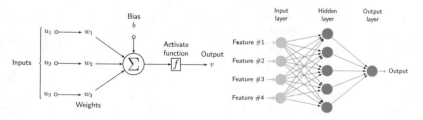

Fig. 11.7 Left: General architecture of a neuron. Right: General architecture of an ANN

where f is the nonlinear activate function (an S-shaped function, classically a logistic or tanh function), u_i are the ith local input features of the neuron (components of $\mathbf{u}$ the local input vector), w_i is the weight attached to u_i and b an added bias.

Given a training data of observed inputs and outputs, and a chosen architecture, the ANN is trained when it learned the optimal weights. The weights are optimal when the local output is as close as possible to the predicted output, at each neuron. A new point prediction is then performed by passing the point input values through the ANN, using the computed weights, until the last output layer. Note that these weights are initialized at the beginning with random values and do not carry any meaning. The backbone of ANNs is therefore the method to update these weights according to the data. One of the most common and powerful algorithms to train the weights is the *backpropagation algorithm* (BP) and its different variants [52]. It is a gradient descent algorithm. It looks to improve the performance of the network by changing the weights along its gradient in order to minimize the total error between observed and predicted output, taken as an RMSE [64]:

$$\text{Error}_{\text{BP}}(\mathbf{w}) = \frac{1}{2}\mathbb{E}\left[\left\{y - y_{out}(\mathbf{x}, \mathbf{w})\right\}^2\right] \qquad (11.28)$$

where $\mathbb{E}$ is a mean function (expectation), y is the desired output of the network, and $y_{out}(\mathbf{x}, \mathbf{w})$ is an output estimated by the ANN for y, with $\mathbf{x}$ the input vector and $\mathbf{w}$ the vector of weights chosen by training. BP initializes the weights randomly. Then, at each neuron, the v output is calculated and an error term for each node is computed backward through the networks, starting from the final output all the way to the input vector. We do not detail here the BP algorithm and the optimization process to choose the weights. These are reviewed extensively in [26].

The architecture of the ANN is crucial (number of neurons, layers, etc.), since it is determinant for the final error on the prediction. One can make this architecture as complex as desired (by adding hidden layers with various numbers of neurons), which makes the ANN a powerful tool to handle large and complex systems with interrelated parameters. However, there are no automatic ways to choose this architecture, which implies the necessity of a methodical and tedious process to test different configurations. In return, the theory guarantees that a multilayer feed-forward (typical type of ANN we presented) ANN can learn almost any pattern, given that the architecture is well chosen and the number of steps large enough [56]. Also, note that ANNs can be used for both classification and regression tasks. The difference between the two tasks resides in the nature of the output we feed the ANN (either an integer or a real number) and the error function used to optimize the weights.

ANNs have been getting a lot of renewed attention over the past few years in the machine learning community. The main reason for that is the available computational power that dramatically increased allowing to build very deep—with a very large number of hidden layers and hidden variables—and complex networks. Examples of such networks are Recursive Neural Networks and Convolutional Neural Networks [71]. This family of very large networks is now referred to as Deep Learning methods and is a very popular and trending topic in ML.

11.7.2.2   Support Vector Machines

Support Vector Machines (SVM) are one of the most common and efficient machine learning algorithms for classification tasks and perhaps the flagship of the kernel methods. It was developed by Cortes and Vapnik in 1995 [30] and improved at various levels through the years. It was also successfully extended to a regression setting, under the name of Support Vector Regression (SVR) [37]. We will present the algorithm for classification, Support Vector Classification (SVC), since it was the original idea of the method. SVR, however, is close to SVC in principle and easily understandable once SVC is grasped. A useful tutorial on SVR can be found in [105].

For simplicity, let us first define a binary classification problem—the principle of the algorithm can be easily extended to a multiclass problem. We consider a set of training data with two classes of points C_1 and C_2. The goal of a binary classification task is to design a function f that can assign any new point $\mathbf{x}$ to either C_1 or C_2, with the help of the training data providing us with examples. The basic idea of SVM is to find the best boundary separating the two classes. The best boundary is found by maximizing the distance between the two classes, called the margin. If the two groups of points are separable by a linear space, we say that they are *linearly separable*. This linear space in 2D is a straight line, a plane in 3D and a so-called *hyperplane* (of dimension $d-1$) in an arbitrary dimension d.

The linearly separable case is fundamental since it develops the basis of the algorithm. Let us write it more formally. We consider the set of N training data points $(\mathbf{x}_n, y_n)_{n=1,\dots,N}$ where each $\mathbf{x}_n$ is a point input vector containing the features of interest and y_n is the corresponding class label, for example -1 for class C_1 or $+1$ for class C_2. We consider the data to be linearly separable, so we know the form of the solution: It is a hyperplane separating the positive from the negative examples. Since it is a hyperplane, it can be parametrized by a vector normal (perpendicular) to the hyperplane and we call this vector $\mathbf{w}$. We also consider the offset b, defining the shift between the hyperplane and the origin of the space where the data points live (the dimension of that space is equal to the number of features). The points that lie on the hyperplane satisfy $\mathbf{w}^\top \mathbf{x} + b = 0$, where $\mathbf{w}^\top \mathbf{x}_n$ is the scalar product between $\mathbf{w}$ and $\mathbf{x}$, in a matrix formulation. The quantity $\frac{|b|}{\|\mathbf{w}\|}$ is the distance from the hyperplane to the origin, where $\|\mathbf{w}\|$ is the Euclidean norm of $\mathbf{w}$, a measure of its length. For the linearly separable case, our f function is then [23]:

$$f(\mathbf{x}) = \text{sign}\left(\mathbf{w}^\top \mathbf{x} + b\right) = \begin{cases} +1 & \text{if } \mathbf{x} \in C_1 \\ -1 & \text{if } \mathbf{x} \in C_2 \end{cases} \tag{11.29}$$

where C_1 and C_2 are two considered classes. The solution is the hyperplane with the highest margin, and the training points satisfy the following constraints:

$$\mathbf{w}^\top \mathbf{x}_n + b \geq +1, \text{ for } y_n = +1 \text{ with } n = 1, 2, \dots, N. \tag{11.30}$$

$$\mathbf{w}^\top \mathbf{x}_n + b \leq -1, \text{ for } y_n = -1 \text{ with } n = 1, 2, \dots, N. \tag{11.31}$$

Fig. 11.8 Illustration of
SVM for the linearly
separable case, in 2D. The
support vectors are
highlighted in red

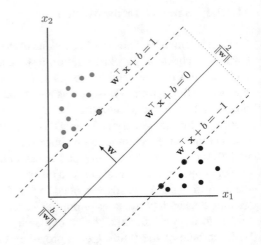

These two set of constraints can be combined in one more convenient one:

$$y_n \left(\mathbf{w}^\mathsf{T} \mathbf{x}_n + b \right) \geq 1, \ n = 1, 2, \dots, N. \tag{11.32}$$

Equation 11.32 will form the constraints of our optimization problem. Now it needs
a clear objective function to be optimized. As it was explained earlier, the objective
is to maximize the margin, the distance between the two classes. A small derivation
leads to a value of $\frac{2}{\|\mathbf{w}\|}$ for this margin, as shown in Fig. 11.8. Since (i) maximizing
a quantity is equivalent to minimizing its inverse, and (ii) manipulating the square
of a norm is equivalent and usually simpler than manipulating the norm itself, the
optimization problem is finally the following [23]:

$$\begin{aligned} \text{Minimize} \quad & \tfrac{1}{2} \|\mathbf{w}\|^2 \\ \text{subject to} \quad & y_n \left(\mathbf{w}^\mathsf{T} \mathbf{x}_n + b \right) \geq 1, \ n = 1, 2, \dots, N. \end{aligned} \tag{11.33}$$

One can use the Lagrangian formulation to solve this optimization problem, which
gives expressions for $\mathbf{w}$ and then f:

$$\mathbf{w} = \sum_{n=1}^{N} y_n \alpha_n \mathbf{x}_n, \ \text{and} \ f(\mathbf{x}) = \text{sign} \left(\sum_{n=1}^{N} y_n \alpha_n \mathbf{x}_n^\mathsf{T} \mathbf{x} + b \right) \tag{11.34}$$

It yields the so-called *dual problem*:

$$\text{Minimize}_{\alpha} \quad D(\alpha) = \frac{1}{2} \sum_{n=1}^{N} \sum_{m=1}^{N} y_n y_m \alpha_n \alpha_m \mathbf{x}_n^\top \mathbf{x}_m - \sum_{n=1}^{N} \alpha_n$$

$$\text{subject to} \quad \alpha_n \geq 0, \ n = 1, 2, \ldots, N,$$

$$\sum_{n=1}^{N} \alpha_n y_n = 0.$$

(11.35)

where α_n are the Lagrange multipliers for the constraints of the problem.

The dual formulation is a quadratic problem that can thus be solved with Quadratic Programming (QP). Solvers for QP give us the α_n, which allows to obtain the final solution for $\mathbf{w}$ and then f. Note that there is one α_n for each training point. The training points corresponding to $\alpha_n \geq 0$ are called the *support vectors* and lie on one of the two hyperplanes defining the boundaries of the classes. Note also that everything is written in terms of scalar products in the dual formulation, which makes it the key for the non linear case, as it will be explained further in the next paragraphs.

It exists another formulation of the linear SVM classification problem that allows the margin to be more flexible, by relaxing the constraints. This formulation, called the *soft-margin* SVM, allows for points slightly outside of a class (inside the margin) to still be considered in this class, with an added penalty. This penalty is expressed by slack variables (one for each training point) ξ_n that measures how far from the boundaries the points are [30]. In order to control the impact of the created penalty on the objective function, a parameter C is added as a multiplier in front of the sum of the slack variables. C determines the trade-off between the generalization of the classifier and the amount of outliers tolerated (if C is too big, the data will be over-fitted and the capacity to adapt to a new data will be low; if C is too small, it might adapt well but will not have enough memory of the training data to classify well). As a parameter, C will later have to be tuned to optimize the model. This soft classification is now formulated as follows:

$$\text{Minimize}_{\mathbf{w}} \quad \frac{1}{2} \|\mathbf{w}\|^2 + C \sum_{n=1}^{N} \xi_n$$

$$\text{subject to} \quad y_n \left(\mathbf{w}^\top \mathbf{x}_n + b \right) \geq 1 - \xi_n, \ n = 1, 2, \ldots, N.$$

(11.36)

The Lagrangian formulation is similar to the previous one, with the extra slack term added. The obtained dual problem is almost the exact same as in Eq. 11.35 with the only difference that the α_n are bounded by C in the first set of constraints: $0 \leq \alpha_n \leq C$, $n = 1, 2, \ldots, N$.

The main contribution of SVM as part of the kernel methods family is that it can also yield very good results in the nonlinear case. In that case, the points are implicitly mapped into a higher dimension space by the so-called *kernel trick*, as illustrated in Fig. 11.9. This trick consists of implicitly applying a nonlinear mapping

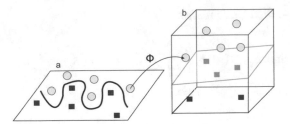

Fig. 11.9 Nonlinear (kernel) SVM principle. Once mapped in a 3D space (the high-dimensional mentioned for Eq. 11.37, in (**b**)), the points are naturally separable by a linear space, here a 2D plane (the feature space, in (**a**)). It translates into a nonlinear line once mapped back into the original 2D space

Φ to the input vectors so that the transformed inputs live in a space where they are linearly separable. This is achieved by expressing scalar products with a nonlinear function called a *kernel*. Yet, the nonlinear mapping Φ is never explicitly chosen, and everything is described by the kernel K [30]:

$$K\left(\mathbf{x}_i, \mathbf{x}_j\right) = \Phi(\mathbf{x}_i)^\top \Phi(\mathbf{x}_j) = \mathbf{z}^\top \mathbf{z}' \tag{11.37}$$

where $\mathbf{z}$ and $\mathbf{z}'$ are living in some higher dimensional space Z.

A crucial point of the kernel trick is that all the computations are performed in the original space through the kernel and do not need to be done in the Z space, where the high dimension would make the calculation potentially intractable. The mathematical formulation and solution remain the same, with the only difference that the scalar product $\mathbf{x}_n^\top \mathbf{x}_m$ is replaced with the kernel $K\left(\mathbf{x}_n, \mathbf{x}_n\right)$ (the program is still solvable by QP). There exists a number of different kernels, satisfying several mathematical conditions in order to be considered proper kernels (the details are not given here). Yet, the most famous one (for its very high efficiency) is the Gaussian kernel, defined by $K\left(x_i, x_j\right) = \exp\left(-\frac{\|x_i - x_j\|^2}{2\sigma^2}\right)$, where σ is a parameter, defining the width of the "bell" that has to be chosen appropriately. Note that the other parameter of SVM (brought by the soft-margin formulation) that needs to be tuned is the trade-off C. Depending on the choice of the kernel, there might be additional parameters. A good review on SVM can be found in [23]. Eventually, the kernel trick is very powerful since it can be (and has been) embedded in all sorts of methods to extend them to a nonlinear setting. That is part of the reason why kernel methods were very popular and extensively studied in the years 2000s.

11.7.2.3 Random Forests

Random Forest (RF) is an algorithm developed by Breiman [19] that allows to perform both classification and regression tasks in a very fast and efficient manner by combining the estimations of multiple classification or regression trees [20]. Such decision trees are very fast but very poor (inaccurate) learners. However, it was discovered that aggregating a large number of poor learners could lead to a very good performance, while keeping a reasonable speed for the training process. This aggregation is characterized as ensemble learning.

Decision trees are models that are trained by a series of binary splits over the training data. Given a collection of feature variables $X_1, X_2, \ldots, X_d$, and our N data points $(\mathbf{x}_1, \mathbf{x}_2, \ldots, \mathbf{x}_N)^{\mathsf{T}}$, the tree iteratively separates the data into subsets of points, according to a query on one of the variables. Such a query can be, for instance, $X_4 > 2.7$, and is chosen with the data according to an optimization process that is the core of the algorithm and presented in the following parts. The query leads to a separation of the current data subset to be further split into two subsets, one for which all the points are characterized by $X_4 > 2.7$ (traditionally pictured as the left one) and one for which all the points are characterized by $X_4 \leq 2.7$ (traditionally pictured as the right one). Each subset is modeled by a *node* of the tree. The first split is operated on the complete data, at the *root node* of the tree, and splits are performed until it is no further possible to split or until a certain chosen depth of the tree has been reached. At each split, the resulting two nodes are called the *children* nodes. The terminal nodes are called *leaves* of the tree. These iterative splits virtually split the input space into multiple subspaces defined by the multiple consecutive queries operated on the variables. Each leaf corresponds to one of these subspaces and is gathering a subset of data points living in this subspace. Once the tree is built with the training data, the prediction of the output value of a new point is performed by passing the point through the tree. The path of the point is defined by the different queries on the point features (at each node, the left path is taken if the query is true, the right path otherwise), until a leaf is reached. This leaf represents the subspace in which the new point lives. In a regression task, the predicted value is then the average of the labels of the training data points gathered in this leaf. In a classification task, the predicted class is the most frequent class label in the training data points gathered in this leaf. Also, note that each node can be characterized by a current estimate, being the average output value of the training samples in this node.

As mentioned earlier, the most important part of the algorithm is the choice of the query to split the data at each node. This choice is done according to the *impurity* metric, defined at each node. The node impurity expresses how far the current node average label is from each of the labels in that node. Since each of the data points in that node would be predicted by the tree with the current average label (the current average label represents the current estimate), the impurity actually measures a current prediction error of the tree. As a result, the impurity needs to be minimized along the tree. Note that the impurity necessarily decreases during the building of the tree (except in some rare cases which not discussed here), since the number of training samples in each node decreases. Consequently, the query choice is done to maximize the impurity decrease $\Delta i(n)$ between the current node and the two children nodes, and can be expressed as [77]:

$$\Delta i(n) = i(n) - \frac{N_{n_L}}{N_n} i(n_L) - \frac{N_{n_R}}{N_n} i(n_R) \tag{11.38}$$

where $i(.)$ denotes the impurity, n the current node, n_L the left child node, n_R the right child node, and N_n, N_{n_L}, and N_{n_R} are, respectively, the number of training data samples in the current node, left child node, and right child node. The impurity of the node, in the regression case, is taken as a squared loss function:

$$i(n) = \frac{1}{N_n} \sum_{j \in \{k | x_k \in n\}} \left(y_j - \bar{y}\right)^2 \tag{11.39}$$

where $j \in \{k \mid x_k \in n\}$ is the set of indices for which the training samples are contained in current node n (the y_j are the output values corresponding to the training samples contained in node n) and $\bar{y}$ is the current node estimate, meaning the average of the y values in the node. In the classification case, the impurity can be expressed by multiple functions, among which the most famous are the ones based on the Gini index and the Shannon entropy [77]. There is at each node an optimization process to choose the best feature and the best threshold to split on. The optimal feature and threshold are obtained in order to maximize the impurity decrease (the details of the optimization in practice are not given here).

Random Forests aim to improve the performance of a single tree estimators by considering multiple trees built with the training data and aggregating their prediction. This aggregation eventually leads to a better accuracy because the variance of the resulting estimator is reduced. In order to build multiple trees using the same training data, RFs fit a large number B of trees to bootstrapped versions of the training data, meaning B data sets of the same size N, created from sampling randomly from the training data with replacement. One issue is the clear correlation between the B resulting trees, which affect the final performance of the forest. In order to de-correlate the trees, RFs add some randomness in the building of each tree: When splitting the node, instead of considering all features as candidates for the best split, only m variables out of all are considered and the best among these m variables is used to split the node. Not only this additional randomness layer allows to considerably increase the performance of the forest but makes it a very fast algorithm. Once all trees have been built, the prediction estimate from the RF is simply the average of the estimations from all trees, in the regression case, and the majority of *votes* from all trees (the most frequent predicted class), in the classification case. An illustration of the training and prediction processes with RFs is shown in Fig. 11.10.

In addition to its speed and great performance, RFs have a number of practical advantages that make the data processing and model selection very easy compared to many other ML algorithms: (1) It is not necessary or advantageous to scale or normalize the data prior to the training process; (2) the optimization embedded at each node of the trees results in automatic feature selection (if some variables are

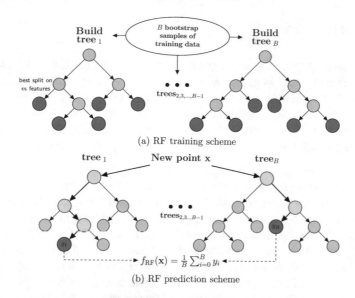

(a) RF training scheme

(b) RF prediction scheme

Fig. 11.10 An illustration of the training and prediction processes with Random Forests, for the regression case. For the classification case, the final prediction is given as the most frequent class predicted from the trees

not informative, the algorithm will not consider them automatically); (3) RFs handle outliers in a way that they do not affect the performance of the prediction, and thus, there is no need to remove them (they may simply result in isolated nodes); (4) a measure of the importance of each variable can be provided [19]; (5) a validation error is already computed during the training, in the form of the so-called Out-Of-Bag (OOB) estimate of error [19]; (6) RFs ultimately have only two significant parameters to tune: B and m, respectively, being the number of bootstrap trees, and the number of variables considered for splitting at each node. Also, the performance can only increase with B, so there is no need to fine-tune it. Consequently, it is perfectly acceptable to fix m and try increasing values of B until an error plateau is reached [50]; (7) even though m needs to be tuned to guarantee optimal results, experience shows that the performance of an RF is not significantly affected by m. Thus, it is current practice to simply try multiple values recommended by Breiman [74] and pick the best one.

Valuable sources on machine learning theory and algorithms can be found in [50, 116]. Another very useful resource on the use of machine learning for environmental data can be found in [64].

11.7.3 Application

Machine learning (ML) has become greatly popular over the last two decades and is now commonly used in many fields, including renewable energy estimation and integration [123]. Since it is solely based on data, it can be used to estimate any kind of variable. The only requirement is a sufficient amount of training data. As a result, ML methods have been suggested to estimate some of the variables of interest for the rooftop PV potential, in several studies. We will now present some of these studies.

ML for radiation mapping. For global solar energy estimation, multiple learning methods have been used in the literature, among which the three presented methods are probably the most popular. ANNs in particular have been extensively used with different input features and network architectures. An example of typical ANN architecture for solar prediction is shown in Fig. 11.11. In [82], an ANN was built, using data from 41 stations, to estimate the global solar radiation for a few zones in Saudi Arabia. The chosen ANN was a multilayer ANN with the backpropagation algorithm as the training algorithm. Its architecture consists of four inputs features, being longitudes, latitudes, altitude, and sunshine duration, ten hidden neurons, and one output neuron, the solar radiation. Its achieved a MAPE error of 12.6%. Another multilayer ANN was used in [2] to estimate the global solar radiation in Oman with more input features: location, mean pressure, month, temperature, vapor pressure, sunshine duration, and relative humidity. The MAPE was 7.9%. In [36], a recurrent neural network (accounting for output feedback) is considered to estimate solar radiation in Cyprus. Backpropagation was used for both these models. In [106], an ANN using location, month, sunshine ratio, and temperature was used to estimate solar radiation for 12 different zones in Turkey. The training is based on the Levenberg-Marquardt learning algorithm and the use of a logistic sigmoid transfer function. The average MAPE is 6.8%; however, only short-term data is used, which does not guarantee a very good accuracy. A feedback propagation neural network is proposed

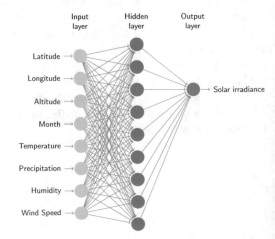

Fig. 11.11 Example of ANN architecture to estimate global solar radiation

in [41] to estimate monthly global solar radiation for Nigeria. In [11], an RBF neural network using RBF functions as activation functions is used to model global solar radiation in Al-Madinah, Saudi Arabia, with temperature, sunshine ratio, and relative humidity as inputs. The performance is significantly improved compared to older models, since the average MAPE is 1.75% . A more comprehensive review on solar prediction with ANN can be found in [125]. In [124], a comparative study for daily solar radiation estimation between regular ANN models, Bayesian Neural Networks (BNN), and empirical models is presented. BNN is modified ANN models in which the learning process is conducted using a Bayesian probabilistic framework. In particular, the weights are chosen based on their estimated conditional probability. The results showed a better performance from the BNN, compared to other models.

Some recent studies propose solar prediction with SVM models, but most of them forecast solar radiance for a very specific place (with historical data on one station) and do not perform spatial interpolation. In [29], SVR is used to forecast future monthly solar radiation for Chongqing meteorological station in China, based on a 28-year collected data in this station. Three kernel functions were tried (linear, polynomial, and RBF kernel), along with different input features (maximum and minimum temperatures, and the difference of both), ending with 21 different SVR models. In this particular study, the polynomial kernels with min and max temperatures showed the best performance. These SVR models were also compared to empirical models and showed general better performances. Daily solar radiation over three stations in Liaoning Province in China was forecast in [27] using sunshine duration. SVR models outperformed the empirical models tested in the study. In [95], different SVR models are tested for the meteorological station in Tehran Province to predict global solar radiation. The input features used for the SVR models are the maximum and minimum temperature, sunshine duration, daylight hour, day number, and extraterrestrial radiation. The results showed that the RBF kernel-based SVR model outperformed the other models and also outperformed other ANN models tried in the study. The transferability of SVR models for global solar radiation is investigated in [28]. The idea is that solar radiation could be well estimated by an SVM model built from another site, depending on the distance and difference of temperature and altitude between the two sites. Yet, the SVM models are still trained on each station separately and tested on the other stations. Location is still not an input for the model. Some spatial generalization is presented in [8], where 14 different Spanish stations are used to build SVR models. Each of the station solar radiation is estimated using an SVR model trained on the other 13 stations. MAE errors averaged 1.81 MJ m^{-2} day^{-1}, which is significantly lower than the errors found on the previous literature with only local models. Assouline et al. [9] suggest the use of SVR directly as a spatial interpolation method to estimate the horizontal global, diffuse, and top of the atmosphere radiations all over Switzerland, using satellite radiation data from 100 different points. The longitude, latitude, altitude, along with weather variables, are considered for training features.

A few recent studies suggested the use of Random Forests for solar radiation estimation. They are, however, less popular than ANN or SVR in this field. This is perhaps because of its smaller exposure to the public outside of the machine learning community. Hashimoto and Nemani [49] use RFs to map surface climate variables, including humidity, wind, and solar radiation. This study is driven by the lack of physical models and governing equations for these variables, in contrast with other well-known climate variables such as temperature and precipitation. The RFs use point data for the three variables as well as satellite and radar data to build the training data. The model is finally applied to the conterminous USA from 1980 to 2015. Sun et al. [109] present an interesting RF study for solar radiation assessment. Its main originality is the use of air pollution index as an extra feature for the RF along with the typical meteorological features. The model is used to forecast the estimate solar radiation in three different sites. The results show that the pollution index improves the model, as it reduces the testing RMSE by 2.0–17.4%. This indicates the potential for such indexes to be used in spatial solar mapping. Kratzenberg et al. [70] use RFs to forecast hourly and daily solar radiation in the southern Brazil region. In a very recent algorithm [57], a hybrid model combining Random Forests with the Firefly algorithm is presented to assess the hourly solar radiation. It showed superiority over traditional RFs and various ANN models.

Besides horizontal radiation, ML algorithms, particularly SVM and ANN, have been used to directly estimate the radiation on a tilted surface, for specific locations avoiding the use of any physical models. Most studies use surface characteristics (slope and aspect), horizontal radiations, and meteorological variables as training features to build their model. Most studies achieve good performance, as, for instance, showed in the work by Gulin et al. [47] using ANNs and a more recent one by Ramli et al. [96]. This last study suggests both ANNs and SVM, and provides a good review of other studies using both algorithms for tilted radiation estimation.

Numerous studies compare the performance of various ML algorithms to estimate solar radiation [42, 118]. The conclusion varies, and the best practice is often to try various algorithms and simply pick the best for the data at hand. However, ML algorithms in general have been used predominantly for forecasting tasks (in one location) and fewer spatial mapping studies are found.

ML for PV suitability variables. The use of ML algorithms to estimate solar PV suitability variables, including the available rooftop area, shadowing effects, slope, aspect, and roof shape, has unfortunately not been nearly as frequent as for the estimation of solar radiation. There are, however, very few studies that use ML to estimate the solar PV suitability variables as presented below. A few studies used ML algorithms to identify roof types, in order to assess the solar potential over rooftops. Joshi et al. [62] present a two-stage classification method to detect rooftops and compute the total roof area with a case study of Abu Dhabi. Satellite images are used in the first stage with an ANN (a multilayer perceptron) and various image features to detect rooftop areas from nonrooftop areas and compute class probabilities for each type of area. In the second stage, the class probabilities from the first stage are used for features, along with shadowing variables. An SVM is trained with these features in order to perform again the classification. It shows an improved accuracy than the one obtained when using the MLP alone. A coefficient of 65% is then used to obtain the available area. The global horizontal radiation is extracted from the Global Solar Radiation (GSR) data from NASA. The technical PV potential for rooftops in Abu Dhabi as well as an economical study are provided. Mohajeri et al. [81] use SVM to classify building roof shapes in the city of Geneva in order to analyze the impact of various roof characteristics on the solar potential in Switzerland. LiDAR data is used to extract aspect and slope features. In addition, a vector polygon building data is used to extract extra geometrical features to characterize the roof shapes. The model identifies six types of roof shapes (flat, gable, hip, gambrel and mansard, cross/corner gable and hip, and complex roof) with a testing accuracy of 66%. Roof types are ranked based on their suitability for PV installations. The flat and hip roofs eventually show, respectively, the highest and lowest potential for solar PV.

In a very recent study, a complete methodology using SVR is suggested by Assouline et al. [9] in order to assess the rooftop PV potential in the urban areas of Switzerland at commune (municipality) level. The study uses solar satellite data and meteorological measurement point data together with SVR to build solar maps for the global, diffuse, and extraterrestrial radiations. A GIS footprint polygon data is used to locate the buildings in urban areas together with the CORINE land cover data. Further GIS processing is performed over a precise vector rooftop database available in the Geneva canton, to extract the average available area for PV, slope distribution, and shadowing effects (using hillshade analysis) over rooftops in the communes of canton (42 communes). These two datasets are used to train and validate SVR models, together with various building characteristics used as training features (population and building densities, footprint area, building typologies, etc.). These trained models are then used in the rest of the communes to estimate the average available area, slope distribution, and shading impact over rooftops. The aspect distribution of the buildings is estimated based on the footprint polygon data. The global tilted radiation is then computed with physical models (Reindl [100] and Klein [67]), combining the estimated solar horizontal radiations with slope, shading and aspect values. Eventually, the estimated available area is used to compute the final technical PV potential in each commune, and the total PV potential in Switzerland is estimated at 17.86 TWh/year (considering a PV efficiency of 17% and a performance ratio of 80%). The obtained technical rooftop PV potential map of Switzerland is shown in Fig. 11.12. Testing RMSE errors are provided for all the estimated variables and show reasonable performances. To the best of our knowledge, it is the first study using machine learning to compute all the necessary variables leading to the rooftop PV potential.

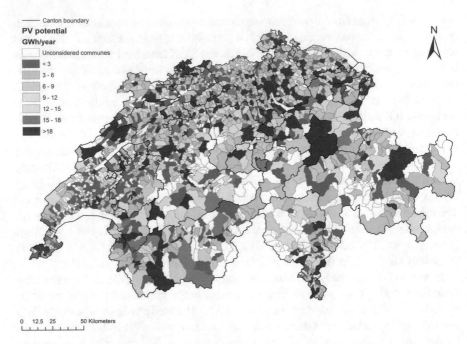

Fig. 11.12 Technical potential of rooftop PV electricity production (GWh/year) for each commune in Switzerland. This figure is reused via Elsevier License #4134701271907, from [9]

11.8 Summary and Conclusion

Some of the most widely used methodologies which lead to the estimation of the rooftop PV potential at a large scale have been presented. For each method, we provide a theoretical background and a literature review of various studies using the method of interest in different contexts.

Six families of methods, useful for various steps in the rooftop potential, were identified:

- Physical and empirical models. They provide governing equations and parametric models for solar horizontal and tilted radiation.
- Geostatistical methods. They gather statistical methods to perform spatial data analysis, particularly prediction and mapping. The most famous and classical family of geostatistical methods is the kriging methods that have been used regularly for solar radiation mapping.
- Constant-value methods. In these methods, simple multiplicative coefficients are used to account for constraints in the potential estimation, such as the available area or shading impact over rooftops for an entire region. These coefficients can be purely assumed or estimated based on various rules of thumb.
- Sampling methods. A variable of interest for a sample of known locations (where precise and adequate data is available) is computed, and the results are extrapo-

lated to the remaining parts of the studied region. They have been used mostly for available area and shading impact calculation.

- GIS and LiDAR data-based methods. They aim to compute the variable of interest in the entire region of interest, using high-resolution aerial photographs or LiDAR data (laser data giving elevation points of the surface of the earth, including buildings, trees). They have been used extensively to compute the available area for PV, shadowing, slope, and aspect of rooftops.
- Machine learning methods. They consist of building models that can learn from examples. Given data, it is therefore possible to build machine learning models to predict any kind of variable, including all the variables of interest for PV potential estimation.

It should be acknowledged that each method has its own advantages and disadvantages, and the user should pick the right method accordingly to the data at hand and the desired level of precision. We provide a summary of the general advantages and disadvantages of each presented method in Table 11.2. The applicability of each method for the various variables of interest is also given in Table 11.3. The key findings of this study for the estimation of rooftop PV potential are the following:

- Concerning the modeling of the tilted radiation over rooftops, the biggest challenge is the estimation of the diffuse tilted radiation. Several models have been

Table 11.2 Summary of methods

Models	Advantages	Disadvantages
Physical/empirical	• Easy to use • Based on governing equations or validated experiments	• Empirical models are not accurate • Based on complex parameters
Geostatistical	• Probabilistic outputs (confidence intervals, risk maps) • Well-established theory	• Computationally intensive • Hard to account for many features • Not extremely scalable
Constant values	• Easy and fast to compute • Easily scalable	• Poor accuracy • Local characteristics discarded • Hard to validate
Sampling	• Based on real data • Scalable and not very expensive • Probabilistic outputs	• Does not assure good extrapolation • Varying performance
GIS/LiDAR	• Detail specific, good accuracy • Can be automated for multiple regions	• Time intensive • Computationally expensive • Hard to scale up
Machine learning	• State-of-the-art performance • Ability to account for many features • Local adaptability • Easily scalable • Very volatile (only adequate data needed)	• Need large amount of data • Possibly hard to train (for some models, such as ANNs)

Table 11.3 Applicability of methods for significant rooftop PV potential variables

Methods	Radiation mapping	Tilted radiation	Available area	Shading	Slope and aspect	Roof type
Physical/empirical	✓	✓				
Geostatistics	✓	✓				
Constant values			✓	✓		
Sampling			✓	✓	✓	✓
GIS/LiDAR	✓		✓	✓	✓	✓
Machine learning	✓	✓	✓	✓	✓	✓

developed to estimate the diffuse tilted radiation based on horizontal components of the radiation and other variables. Depending on the local weather conditions, the different models can offer a varying performance and thus provide different results. As a result, when computing the diffuse radiation over a rooftop in a location of interest, it is advised to try a variety of diffuse models and ideally test their performance with local measured data.

- Concerning radiation mapping, geostatistics (particularly the family of kriging models) is a very useful tool with a well-established theory. They provide probabilistic outputs, leading to a measure of the errors attached to the estimates. They require, however, extensive computations. Machine learning algorithms can be a good alternative to perform spatial interpolation, particularly when a significant amount of measurement data is available.
- In order to estimate various rooftop characteristics, constant-value methods lack in precision and are only suitable to compute an indication of the PV potential for very large areas or extract a first approximation of the potential. They are useful when very few data are available. They should not, however, be used in the context of a precise study leading to an actual implementation plan.
- When precise data is available solely in a few specific locations, sampling methods are suitable to compute various rooftop characteristics, particularly for large-scale studies. Nevertheless, they require a thorough statistical validation of the samples and the extrapolation strategy in order to provide accurate results.
- In recent years, GIS and LiDAR data-based methods have been predominant due to the considerable increase in the availability of LiDAR data everywhere in the world. As they also allow automation of computation and provide satisfying results (specially if some very high-resolution LiDAR data is available), they are often considered to be the best methods to estimate the multiple variables relative to the rooftop PV potential. They require, however, very heavy computer calculations to provide accurate results. Therefore, they are particularly suitable for small to medium scale studies (e.g., at a neighborhood or city level) but are troublesome to scale to large regions, such as whole countries.
- Machine learning methods can be a good alternative to LiDAR-based methods that are computationally very demanding, to compute all variables leading to the

rooftop PV potential. ML methods are conceptually equivalent to advanced sampling methods and do not necessarily require tremendous computational power, but solely representative data to learn from. Furthermore, machine learning is a rapidly growing field and is being used in countless domains to take advantage of the massive amount of data available.

Renewable energy potential estimation, in particular for rooftop PV electricity, is a relatively new field and is expected to benefit from significant research in the next few years. Such future works include:

- The use of very high-resolution data to compute the solar potential for PV installation not only over rooftops but also over building facades, at a large scale.
- The use of advanced image processing techniques combined with satellite imagery to provide an accurate and less computationally demanding alternative to LiDAR-based methods. Such studies [62] have been proposed only at relatively small scales, mainly because high-resolution LiDAR data is not always available over a large territory. The availability of such data is fortunately increasing.
- The integration of the PV electricity and other renewable energies potential studies in smart grids and future power systems. Several studies have proposed smart strategies for the management of distributed renewable resources [4, 17]. These strategies can be embedded within potential studies in order to provide the potential for multiple resources of energy within a central network.
- The development of machine learning strategies to compute the potential for other possible renewable sources of energy in urban environments such as geothermal heat pumps or wind turbines.
- The development of methodologies to account for the economical and market-related constraints over the rooftop PV potential. They rely on very different methods, in nature, than the ones presented here, and are to be specifically adapted to each particular location. They were consequently discarded in this particular review. Nevertheless, they represent a fundamental aspect of the PV potential and need to be accounted to provide a reliable estimation of the actual electricity output from the PV panels to be installed over rooftops.

Acknowledgements This research has been financed partly by the Commission for Technology and Innovation (CTI) within the SCCER Future Energy Efficient Buildings and Districts, FEEB&D (CTI.2014.0119) and partly by Swiss National Science Foundation under Mobility Fellowship P300P2 174514.

References

1. Y.A. Abdalla, M. Baghdady, Global and diffuse solar radiation in Doha (Qatar). Solar Wind Technol. **2**(3–4), 209–212 (1985)
2. S. Al-Alawi, H. Al-Hinai, An ANN-based approach for predicting global radiation in locations with no direct measurement instrumentation. Renew. Energy **14**(1–4), 199–204 (1998)
3. H. Alsamamra, J.A. Ruiz-Arias, D. Pozo-Vázquez, J. Tovar-Pescador, A comparative study of ordinary and residual kriging techniques for mapping global solar radiation over southern Spain. Agric. Forest Meteorol. **149**(8), 1343–1357 (2009)

4. M.H. Amini, B. Nabi, M.-R. Haghifam, Load management using multi-agent systems in smart distribution network, in *2013 IEEE Power and Energy Society General Meeting (PES)* (IEEE, 2013), pp. 1–5

5. P. Andersen, Comments on calculations of monthly average insolation on tilted surfaces by SA Klein. Solar Energy **25**(3), 287 (1980)

6. K.H. Anderson, M.H. Coddington, B.D. Kroposki, Assessing technical potential for city PV deployment using NREL's in my backyard tool, in *2010 35th IEEE Photovoltaic Specialists Conference (PVSC)* (IEEE, 2010), pp. 001085–001090

7. A. Angstrom, On the computation of global radiation from records of sunshine

8. F. Antonanzas-Torres, R. Urraca, J. Antonanzas, J. Fernandez-Ceniceros, F. Martinez-de Pison, Generation of daily global solar irradiation with support vector machines for regression. Energy Convers. Manag. **96**, 277–286 (2015)

9. D. Assouline, N. Mohajeri, J.-L. Scartezzini, Quantifying rooftop photovoltaic solar energy potential: a machine learning approach. Solar Energy **141**, 278–296 (2017)

10. V. Badescu, 3D isotropic approximation for solar diffuse irradiance on tilted surfaces. Renew. Energy **26**(2), 221–233 (2002)

11. M. Benghanem, A. Mellit, Radial Basis Function Network-based prediction of global solar radiation data: application for sizing of a stand-alone photovoltaic system at Al-Madinah, Saudi Arabia. Energy **35**(9), 3751–3762 (2010)

12. R. Benson, M. Paris, J. Sherry, C. Justus, Estimation of daily and monthly direct, diffuse and global solar radiation from sunshine duration measurements. Solar Energy **32**(4), 523–535 (1984)

13. L. Bergamasco, P. Asinari, Scalable methodology for the photovoltaic solar energy potential assessment based on available roof surface area: application to Piedmont Region (Italy). Solar Energy **85**(5), 1041–1055 (2011)

14. L. Bergamasco, P. Asinari, Scalable methodology for the photovoltaic solar energy potential assessment based on available roof surface area: further improvements by ortho-image analysis and application to Turin (Italy). Solar Energy **85**(11), 2741–2756 (2011)

15. A. Bill, N. Mohajeri, J.-L. Scartezzini, 3D model for solar energy potential on buildings from urban LiDAR data. UDMV 2016 - Eurographics Workshop on Urban Data Modelling and Visualisation, Liège, Belgium, December 8, 2016

16. A. Bocca, L. Bottaccioli, E. Chiavazzo, M. Fasano, A. Macii, P. Asinari, Estimating photovoltaic energy potential from a minimal set of randomly sampled data. Renew. Energy **97**, 457–467 (2016)

17. K.G. Boroojeni, M.H. Amini, A. Nejadpak, S. Iyengar, B. Hoseinzadeh, C.L. Bak, A theoretical bilevel control scheme for power networks with large-scale penetration of distributed renewable resources, in *2016 IEEE International Conference on Electro Information Technology (EIT)* (IEEE, 2016), pp. 0510–0515

18. M.B. Boz, K. Calvert, J.R. Brownson, An automated model for rooftop PV systems assessment in ArcGIS using LIDAR. AIMS Energy **3**(3), 401–420 (2015)

19. L. Breiman, Random forests. Mach. Learn. **45**(1), 5–32 (2001)

20. L. Breiman, J. Friedman, C. Stone, R. Olshen, Classification and Regression Trees. The Wadsworth and Brooks-Cole Statistics-probability Series (Taylor & Francis, 1984)

21. M.C. Brito, N. Gomes, T. Santos, J.A. Tenedório, Photovoltaic potential in a Lisbon suburb using LiDAR data. Solar Energy **86**(1), 283–288 (2012)

22. J. Bugler, The determination of hourly insolation on an inclined plane using a diffuse irradiance model based on hourly measured global horizontal insolation. Solar Energy **19**(5), 477–491 (1977)

23. C.J. Burges, A tutorial on support vector machines for pattern recognition. Data Min. Knowl. Discov. **2**(2), 121–167 (1998)

24. C. Carneiro, E. Morello, G. Desthieux, Assessment of solar irradiance on the urban fabric for the production of renewable energy using LIDAR data and image processing techniques, in *Advances in GIScience* (Springer, 2009), pp. 83–112

25. M. Chaudhari, L. Frantzis, T.E. Hoff, *PV Grid Connected Market Potential Under a Cost Breakthrough Scenario*, Retrieved on 16 September 2010 (Navigant Consulting, Inc., 2004)
26. Y. Chauvin, D.E. Rumelhart, *Backpropagation: Theory, Architectures, and Applications* (Psychology Press, 1995)
27. J.-L. Chen, G.-S. Li, Evaluation of support vector machine for estimation of solar radiation from measured meteorological variables. Theoret. Appl. Climatol. **115**(3–4), 627–638 (2013)
28. J.-L. Chen, G.-S. Li, B.-B. Xiao, Z.-F. Wen, M.-Q. Lv, C.-D. Chen, Y. Jiang, X.-X. Wang, S.-J. Wu, Assessing the transferability of support vector machine model for estimation of global solar radiation from air temperature. Energy Convers. Manag. **89**, 318–329 (2015)
29. J.-L. Chen, H.-B. Liu, W. Wu, D.-T. Xie, Estimation of monthly solar radiation from measured temperatures using support vector machines—a case study. Renew. Energy **36**(1), 413–420 (2011)
30. C. Cortes, V. Vapnik, Support-vector networks. Mach. Learn. **20**(3), 273–297 (1995)
31. V. D'Agostino, A. Zelenka, Supplementing solar radiation network data by co-Kriging with satellite images. Int. J. Climatol. **12**(7), 749–761 (1992)
32. P. Defaix, W. Van Sark, E. Worrell, E. de Visser, Technical potential for photovoltaics on buildings in the EU-27. Solar Energy **86**(9), 2644–2653 (2012)
33. C. Demain, M. Journée, C. Bertrand, Evaluation of different models to estimate the global solar radiation on inclined surfaces. Renew. Energy **50**, 710–721 (2013)
34. P. Denholm, R. Margolis, Supply curves for rooftop solar PV-generated electricity for the United States. 2008
35. M. Diez-Mediavilla, A. De Miguel, J. Bilbao, Measurement and comparison of diffuse solar irradiance models on inclined surfaces in Valladolid (Spain). Energy Convers. Manag. **46**(13), 2075–2092 (2005)
36. A.S. Dorvlo, J.A. Jervase, A. Al-Lawati, Solar radiation estimation using artificial neural networks. Appl. Energy **71**(4), 307–319 (2002)
37. H. Drucker, C.J. Burges, L. Kaufman, A.J. Smola, V. Vapnik, Support vector regression machines, in *Advances in Neural Information Processing Systems* (1997), pp. 155–161
38. J.A. Duffie, W.A. Beckman, *Solar Engineering of Thermal Processes* (1980)
39. A. El-Sebaii, F. Al-Hazmi, A. Al-Ghamdi, S.J. Yaghmour, Global, direct and diffuse solar radiation on horizontal and tilted surfaces in Jeddah, Saudi Arabia. Appl. Energy **87**(2), 568–576 (2010)
40. C. Ertekin, F. Evrendilek, Spatio-temporal modeling of global solar radiation dynamics as a function of sunshine duration for Turkey. Agric. Forest Meteorol. **145**(1–2), 36–47 (2007)
41. D. Fadare, Modelling of solar energy potential in Nigeria using an artificial neural network model. Appl. Energy **86**(9), 1410–1422 (2009)
42. Y. Feng, N. Cui, Q. Zhang, L. Zhao, D. Gong, Comparison of artificial intelligence and empirical models for estimation of daily diffuse solar radiation in North China plain. Int. J. Hydrogen Energy **42**(21), 14418–14428 (2017)
43. L. Frantzis, S. Graham, J. Paidipati, California rooftop photovoltaic (PV) resource assessment and growth potential by county. Navigant Consulting, California Energy Commission PIER Final Project Report CEC-500-2007-048 (2007)
44. P. Gagnon, R. Margolis, J. Melius, C. Phillips, R. Elmore, Rooftop solar photovoltaic technical potential in the United States: a detailed assessment. Technical report (National Renewable Energy Laboratory (NREL), 2016)
45. J. Gooding, R. Crook, A.S. Tomlin, Modelling of roof geometries from low-resolution LiDAR data for city-scale solar energy applications using a neighbouring buildings method. Appl. Energy **148**, 93–104 (2015)
46. C. Gueymard, An anisotropic solar irradiance model for tilted surfaces and its comparison with selected engineering algorithms. Solar Energy **38**(5), 367–386 (1987)
47. M. Gulin, M. Vašak, M. Baotic, Estimation of the global solar irradiance on tilted surfaces, in *17th International Conference on Electrical Drives and Power Electronics (EDPE 2013)* (2013), pp. 334–339

48. M. Gutschner, S. Nowak, D. Ruoss, P. Toggweiler, T. Schoen, Potential for building integrated photovoltaics. IEA-PVPS Task **7** (2002)
49. H. Hashimoto, R. Nemani, Estimation of spatial variability in humidity, wind, and solar radiation using the random forest algorithm for the conterminous USA, in *AGU Fall Meeting Abstracts* (2015)
50. T. Hastic, R. Tibshirani, J. Friedman, T. Hastie, J. Friedman, R. Tibshirani, *The Elements of Statistical Learning*, vol. 2 (Springer, 2009)
51. J.E. Hay, Calculation of monthly mean solar radiation for horizontal and inclined surfaces. Solar Energy **23**(4), 301–307 (1979)
52. S.S. Haykin, *Neural Networks: A Comprehensive Foundation* (Tsinghua University Press, 2001)
53. J. Hofierka, J. Kaňuk, Assessment of photovoltaic potential in urban areas using open-source solar radiation tools. Renew. Energy **34**(10), 2206–2214 (2009)
54. T. Hong, M. Lee, C. Koo, K. Jeong, J. Kim, Development of a method for estimating the rooftop solar photovoltaic (PV) potential by analyzing the available rooftop area using hillshade analysis. Appl. Energy **194**, 320–332 (2017)
55. M.M. Hoogwijk, On the global and regional potential of renewable energy sources. Ph.D. thesis (2004)
56. K. Hornik, M. Stinchcombe, H. White, Multilayer feedforward networks are universal approximators. Neural Netw. **2**(5), 359–366 (1989)
57. I.A. Ibrahim, T. Khatib, A novel hybrid model for hourly global solar radiation prediction using random forests technique and firefly algorithm. Energy Convers. Manag. **138**, 413–425 (2017)
58. T. Inoue, T. Sasaki, T. Washio, Spatio-temporal Kriging of solar radiation incorporating direction and speed of cloud movementm, in *The 26th Annual Conference of the Japanese Society for Artificial Intelligence* (2012)
59. M. Iqbal, *An Introduction to Solar Radiation* (Elsevier, 2012)
60. S. Izquierdo, M. Rodrigues, N. Fueyo, A method for estimating the geographical distribution of the available roof surface area for large-scale photovoltaic energy-potential evaluations. Solar Energy **82**(10), 929–939 (2008)
61. A. Jochem, B. Höfle, M. Rutzinger, N. Pfeifer, Automatic roof plane detection and analysis in airborne LIDAR point clouds for solar potential assessment. Sensors **9**(7), 5241–5262 (2009)
62. B. Joshi, B. Hayk, A. Al-Hinai, W.L. Woon, Rooftop detection for planning of solar PV deployment: a case study in Abu Dhabi, in *International Workshop on Data Analytics for Renewable Energy Integration* (Springer, 2014), pp. 137–149
63. G.A. Kamali, I. Moradi, A. Khalili, Estimating solar radiation on tilted surfaces with various orientations: a study case in Karaj (Iran). Theoret. Appl. Climatol. **84**(4), 235–241 (2006)
64. M. Kanevski, V. Timonin, A. Pozdnukhov, *Machine Learning for Spatial Environmental Data: Theory, Applications, and Software* (CRC Press, 2009)
65. R. Kassner, W. Koppe, T. Schüttenberg, G. Bareth, Analysis of the solar potential of roofs by using official LIDAR data, in *Proceedings of the International Society for Photogrammetry, Remote Sensing and Spatial Information Sciences (ISPRS Congress)* (2008), pp. 399–404
66. T. Khatib, A. Mohamed, M. Mahmoud, K. Sopian, Modeling of daily solar energy on a horizontal surface for five main sites in Malaysia. Int. J. Green Energy **8**(8), 795–819 (2011)
67. S. Klein, Calculation of monthly average insolation on tilted surfaces. Solar Energy **19**(4), 325–329 (1977)
68. T.M. Klucher, Evaluation of models to predict insolation on tilted surfaces. Solar Energy **23**(2), 111–114 (1979)
69. P.S. Koronakis, On the choice of the angle of tilt for south facing solar collectors in the Athens basin area. Solar Energy **36**(3), 217–225 (1986)
70. M. Kratzenberg, H.H. Zürn, P.P. Revheim, H.G. Beyer, Identification and handling of critical irradiance forecast errors using a random forest scheme—a case study for southern Brazil. Energy Procedia **76**, 207–215 (2015)

71. Y. LeCun, Y. Bengio et al., Convolutional networks for images, speech, and time series, in *The Handbook of Brain Theory and Neural Networks*, vol. 3361, no. 10 (1995)
72. H. Lehmann, S. Peter, *Assessment of Roof and Façade Potentials for Solar Use in Europe* (Institute for Sustainable Solutions and Innovations (ISUSI), Aachen, Germany, 2003)
73. R. Levinson, H. Akbari, M. Pomerantz, S. Gupta, Solar access of residential rooftops in four California cities. Solar Energy **83**(12), 2120–2135 (2009)
74. A. Liaw, M. Wiener, Classification and regression by randomforest. R News **2**(3), 18–22 (2002)
75. B. Liu, R. Jordan, Daily insolation on surfaces tilted towards equator. ASHRAE J. (United States) **10** (1961)
76. A. Lopez, B. Roberts, D. Heimiller, N. Blair, G. Porro, US renewable energy technical potentials: a GIS-based analysis. Technical report, NREL (2012)
77. G. Louppe, Understanding random forests: from theory to practice, arXiv:1407.7502 (2014)
78. P. Loutzenhiser, H. Manz, C. Felsmann, P. Strachan, T. Frank, G. Maxwell, Empirical validation of models to compute solar irradiance on inclined surfaces for building energy simulation. Solar Energy **81**(2), 254–267 (2007)
79. N. Lukač, S. Seme, D. Žlaus, G. Štumberger, B. Žalik, Buildings roofs photovoltaic potential assessment based on LiDAR (light detection and ranging) data. Energy **66**, 598–609 (2014)
80. W.S. McCulloch, W. Pitts, A logical calculus of the ideas immanent in nervous activity. Bull. Math. Biophys. **5**(4), 115–133 (1943)
81. N. Mohajeri, D. Assouline, B. Guiboud, J.-L. Scartezzini, Does roof shape matter? solar photovoltaic (PV) integration on building roofs, in *Proceedings of the International Conference on Sustainable Built Environment (SBE)*, number EPFL-CONF-220006 (2016)
82. M. Mohandes, S. Rehman, T. Halawani, Estimation of global solar radiation using artificial neural networks. Renew. Energy **14**(1–4), 179–184 (1998)
83. M. Montavon, J.-L. Scartezzini, R. Compagnon, Comparison of the solar energy utilisation potential of different urban environments, in *PLEA 2004 Conference* (2004)
84. A. Moreno, M. Gilabert, B. Martínez, Mapping daily global solar irradiation over Spain: a comparative study of selected approaches. Solar Energy **85**(9), 2072–2084 (2011)
85. T. Muneer, Solar radiation model for Europe. Build. Serv. Eng. Res. Technol. **11**(4), 153–163 (1990)
86. H.T. Nguyen, J.M. Pearce, Incorporating shading losses in solar photovoltaic potential assessment at the municipal scale. Solar Energy **86**(5), 1245–1260 (2012)
87. H.T. Nguyen, J.M. Pearce, Automated quantification of solar photovoltaic potential in cities. Int. Rev. Spat. Plan. Sustain. Dev. **1**(1), 49–60 (2013)
88. H.T. Nguyen, J.M. Pearce, R. Harrap, G. Barber, The application of LiDAR to assessment of rooftop solar photovoltaic deployment potential in a municipal district unit. Sensors **12**(4), 4534–4558 (2012)
89. A.M. Noorian, I. Moradi, G.A. Kamali, Evaluation of 12 models to estimate hourly diffuse irradiation on inclined surfaces. Renew. Energy **33**(6), 1406–1412 (2008)
90. J. Ordóñez, E. Jadraque, J. Alegre, G. Martínez, Analysis of the photovoltaic solar energy capacity of residential rooftops in Andalusia (Spain). Renew. Sustain. Energy Rev. **14**(7), 2122–2130 (2010)
91. J. Paidipati, L. Frantzis, H. Sawyer, A. Kurrasch, *Rooftop Photovoltaics Market Penetration Scenarios* (Navigant Consulting, Inc., for NREL, 2008)
92. J. Peng, L. Lu, Investigation on the development potential of rooftop pv system in Hong Kong and its environmental benefits. Renew. Sustain. Energy Rev. **27**, 149–162 (2013)
93. R. Perez, R. Seals, P. Ineichen, R. Stewart, D. Menicucci, A new simplified version of the Perez diffuse irradiance model for tilted surfaces. Solar Energy **39**(3), 221–231 (1987)
94. I.R. Pillai, R. Banerjee, Methodology for estimation of potential for solar water heating in a target area. Solar Energy **81**(2), 162–172 (2007)
95. Z. Ramedani, M. Omid, A. Keyhani, S. Shamshirband, B. Khoshnevisan, Potential of radial basis function based support vector regression for global solar radiation prediction. Renew. Sustain. Energy Rev. **39**, 1005–1011 (2014)

96. M.A. Ramli, S. Twaha, Y.A. Al-Turki, Investigating the performance of support vector machine and artificial neural networks in predicting solar radiation on a tilted surface: Saudi Arabia case study. Energy Convers. Manag. **105**, 442–452 (2015)
97. C. Ratti, N. Baker, K. Steemers, Energy consumption and urban texture. Energy Build. **37**(7), 762–776 (2005)
98. P. Rcdwcik, C. Catita, M. Brito, Solar energy potential on roofs and facades in an urban landscape. Solar Energy **97**, 332–341 (2013)
99. S. Rehman, S.G. Ghori, Spatial estimation of global solar radiation using geostatistics. Renew. Energy **21**(3–4), 583–605 (2000)
100. D. Reindl, W. Beckman, J. Duffie, Evaluation of hourly tilted surface radiation models. Solar Energy **45**(1), 9–17 (1990)
101. L.R. Rodríguez, E. Duminil, J.S. Ramos, U. Eicker, Assessment of the photovoltaic potential at urban level based on 3D city models: a case study and new methodological approach. Solar Energy **146**, 264–275 (2017)
102. Z. Scen, E. Tan, Simple models of solar radiation data for northwestern part of Turkey. Energy Convers. Manag. **42**(5), 587–598 (2001)
103. J. Schallenberg-Rodríguez, Photovoltaic techno-economical potential on roofs in regions and islands: the case of the Canary Islands. Methodological review and methodology proposal. Renew. Sustain. Energy Rev. **20**, 219–239 (2013)
104. A. Skartveit, J.A. Olseth, Modelling slope irradiance at high latitudes. Solar Energy **36**(4), 333–344 (1986)
105. A.J. Smola, B. Schölkopf, A tutorial on support vector regression. Stat. Comput. **14**(3), 199–222 (2004)
106. A. Sözen, E. Arcaklioğlu, M. Özalp, E. Kanit, Use of artificial neural networks for mapping of solar potential in Turkey. Appl. Energy **77**(3), 273–286 (2004)
107. C. Spitters, H. Toussaint, J. Goudriaan, Separating the diffuse and direct component of global radiation and its implications for modeling canopy photosynthesis. Part I. Components of incoming radiation. Agric. Forest Meteorol. **38**(1–3), 217–229 (1986)
108. M. Stone, Cross-validatory choice and assessment of statistical predictions. J. R. Stat. Soc. Ser. B (Methodol.) 111–147 (1974)
109. H. Sun, D. Gui, B. Yan, Y. Liu, W. Liao, Y. Zhu, C. Lu, N. Zhao, Assessing the potential of random forest method for estimating solar radiation using air pollution index. Energy Convers. Manag. **119**, 121–129 (2016)
110. M. Šúri, J. Hofierka, A new GIS-based solar radiation model and its application to photovoltaic assessments. Trans. GIS **8**(2), 175–190 (2004)
111. R.C. Temps, K. Coulson, Solar radiation incident upon slopes of different orientations. Solar Energy **19**(2), 179–184 (1977)
112. A. Tereci, D. Schneider, D. Kesten, A. Strzalka, U. Eicker, Energy saving potential and economical analysis of solar systems in the urban quarter Scharnhauser Park, in *ISES Solar World Congress 2009*, 11–14 Oct 2009, Johannesburg, South Africa (2009)
113. Y. Tian, R. Davies-Colley, P. Gong, B. Thorrold, Estimating solar radiation on slopes of arbitrary aspect. Agric. Forest Meteorol. **109**(1), 67–74 (2001)
114. S. Topçu, S. Dilmaç, Z. Aslan, Study of hourly solar radiation data in Istanbul. Renew. Energy **6**(2), 171–174 (1995)
115. A.J.M. van Wijk, J.P. Coelingh, *Wind Power Potential in the OECD Countries* (Department of Science, Technology and Society, Utrecht University, 1993)
116. V. Vapnik, *The Nature of Statistical Learning Theory* (Springer Science & Business Media, 2000)
117. R. Vardimon, Assessment of the potential for distributed photovoltaic electricity production in Israel. Renew. Energy **36**(2), 591–594 (2011)
118. C. Voyant, G. Notton, S. Kalogirou, M.-L. Nivet, C. Paoli, F. Motte, A. Fouilloy, Machine learning methods for solar radiation forecasting: a review. Renew. Energy **105**, 569–582 (2017)
119. R. Webster, M. Oliver, *Geostatistics for Environmental Scientists* (2001)

120. L. Wiginton, H.T. Nguyen, J.M. Pearce, Quantifying rooftop solar photovoltaic potential for regional renewable energy policy. Comput. Environ. Urban Syst. **34**(4), 345–357 (2010)
121. C.J. Willmott, On the climatic optimization of the tilt and azimuth of flat-plate solar collectors. Solar Energy **28**(3), 205–216 (1982)
122. C.J. Willmott, K. Matsuura, Advantages of the mean absolute error (MAE) over the root mean square error (RMSE) in assessing average model performance. Clim. Res. **30**(1), 79–82 (2005)
123. W.L. Woon, Z. Aung, O. Kramer, S. Madnick, in *Data Analysis for Renewable Energy Integration: 4th ECML PKDD Workshop, DARE 2016*, Riva del Garda, Italy, 23 Sept 2016, Revised Selected Papers, vol. 10097 (Springer, 2017)
124. R. Yacef, M. Benghanem, A. Mellit, Prediction of daily global solar irradiation data using Bayesian neural network: a comparative study. Renew. Energy **48**, 146–154 (2012)
125. A.K. Yadav, S. Chandel, Solar radiation prediction using Artificial Neural Network techniques: a review. Renew. Sustain. Energy Rev. **33**, 772–781 (2014)
126. W. Yao, Z. Li, Y. Lu, F. Jiang, C. Li, New models for separating hourly diffuse and direct components of global solar radiation, in *Proceedings of the 8th International Symposium on Heating, Ventilation and Air Conditioning* (Springer, 2014), pp. 653–663
127. J.K. Yohanna, I.N. Itodo, V.I. Umogbai, A model for determining the global solar radiation for Makurdi, Nigeria. Renew. Energy **36**(7), 1989–1992 (2011)
128. C.-D. Yue, S.-S. Wang, GIS-based evaluation of multifarious local renewable energy sources: a case study of the Chigu area of southwestern Taiwan. Energy Policy **34**(6), 730–742 (2006)

Chapter 12
Optimal SVC Allocation in Power Systems for Loss Minimization and Voltage Deviation Reduction

M. Hadi Amini, Rupamathi Jaddivada, Bakhtyar Hoseinzadeh, Sakshi Mishra and Mostafa Rezaei Mozafar

12.1 Introduction

The complexity of the power systems network is increasing every day owing to both technological and policy changes. More and more unpredictable renewable energy is being injected into the grid which can have detrimental effects on the grid. With the deregulation in place, there is always some imbalance in the supply and demand of the energy that is affecting the frequency of the grid. In addition, the recent advancements pertaining to the smart grid technology, induction of electric vehicles to take active part in the grid, demand side management, etc., have led to grid in-stability.

M. H. Amini (✉)
Department of Electrical and Computer Engineering, Carnegie Mellon University, Pittsburgh, PA 15213, USA
e-mail: amini@cmu.edu

R. Jaddivada
Department of Electrical Engineering and Computer Sciences, Massachusetts Institute of Technology, Cambridge, MA 02139, USA
e-mail: rjaddiva@mit.edu

B. Hoseinzadeh
Department of Energy Technology, Aalborg University, Aalborg, Denmark
e-mail: bho@et.aau.dk

S. Mishra
American Electric Power, Tulsa, OK 74119, USA
e-mail: sakshi.m@outlook.com

M. R. Mozafar
Faculty of Engineering, Department of Electrical Engineering, Islamic Azad University, Hamedan, Iran
e-mail: m.rezaeimozafar@iauh.ac.ir

© Springer International Publishing AG, part of Springer Nature 2018
M. H. Amini et al. (eds.), *Sustainable Interdependent Networks*,
Studies in Systems, Decision and Control 145,
https://doi.org/10.1007/978-3-319-74412-4_12

Conventionally, the system operators have been carrying out the economic dispatch and unit commitment to schedule generators. They simulate disturbances to find out if the solution found previously maintains the grid in place. If any of the constraints fail, they run the optimization problem by limiting the generation values. The result is that the entire power system runs at sub-optimal point, which means that the system is running less efficiently. So, it has always been a trade-off between efficiency and stability margin. The other methods the operators follow in order to ensure the grid stability is to build additional transmission lines and have additional units that can be used as reserves. These involve high investments. In addition, these reserves operate only for a short time period in the event of contingency. In such a scenario, much more efficient and an inexpensive method is to have controllers that could react to the disturbances and restore the normal grid operations.

We have been using excitation controllers for long time; such as automatic voltage regulators (AVRs) and power system stabilizers (PSS). But these controllers are not only slow but also are far from the disturbance which renders them useless when there is a requirement of maintaining transient stability. In recent years, technological advancements in power electronics facilitated development of equipment that could be switched electronically with the ability to handle large amounts of power. These devices are called Flexible AC Transmission System (FACTS) devices. In the event of small disturbances, these prove to be much more efficient and inexpensive. FACTS technologies offer competitive solutions in terms of power flow transfer capability, voltage security, minimization of power losses, improving system damping, etc. There are two categories of FACTS controllers. Each of these devices is discussed in greater detail in [1]: *Series Compensation* (the device is connected in series with the power systems and it works as a controllable voltage source), *Shunt Compensation* (the device is placed in parallel to the power system and it acts as a controllable current source).

Power electronic devices have been widely used to transiently stabilize the power systems. The FACTS devices are example of such devices which could control the flow of power within fraction of seconds. They have the capability to inject or store large amounts of reactive power for short durations, till the disturbance lasts. These fast acting switches when properly controlled using linear control or nonlinear control techniques [2] can ensure the grid stability in the event of small disturbances that last for short durations. However, these devices do not have the capability to inject real power into the grid, which is why they cannot handle severe contingencies such as loss of generator. In such a scenario, we could consider the installation of fast acting storage devices such as flywheel [3]. Resiliency of the power networks plays a crucial role in the proper operation of distributed power systems [4–6]. Several distributed algorithms have been proposed in the literature to deal with the large-scale optimization problems at different layers of power systems, i.e., distribution and transmission levels [7, 8]. Further, FACTS devices can be deployed to enable the fast load management schemes in power distribution networks and microgrids [9–12].

The optimum location and sizing of SVC are a complex mixed-integer non-convex problem owing to the power flow constraints which are not convex. Also, the variable, location of SVC takes integer values. Hence, conventional optimization techniques cannot solve this problem. Numerous techniques for solving the optimal capacitor placement problem in power systems have been reported. These may be classified into the categories: analytic, heuristic, numerical programming and artificial intelligence-based techniques. Heuristic techniques have been applied widely in solving this kind of problem. In order to determine the optimum location of SVC in power system different techniques such as particle swarm optimization (PSO) and genetic algorithm (GA) can be used, as well as new techniques such as novel global harmony search [13], firefly algorithm [14] under study.

12.2 Static Var Compensator

Among the reactive power compensation devices, shunt compensators play a major role in controlling the flow of reactive power to the power network, thereby affecting the system voltage and stability. The SVC is the majorly used shunt compensator device owing to the low cost compared to its counterpart: STATCOM [15]. It is a shunt-connected static VAR generator or absorber with an adjustable output, which allows the exchange of capacitive or inductive current so as to provide voltage support. When installed at proper location, they can also reduce power losses. The importance of selecting an SVC of suitable rating is discussed in [16].

There are two popular configurations of SVC. One is fixed capacitor (FC) and thyristor-controlled reactor model (TCR), and the other is thyristor-switched capacitor (TSC) and TCR configuration. In the limit of minimum and maximum susceptance, SVC behaves like an FC or an inductor. Figures 12.1 and 12.2 show the configuration and V-I characteristics of the SVC, respectively.

A. *SVC Modeling*

There have been numerous SVC models for different kinds of power system study in the literature. Kueck et al. [17] present the SVC total susceptance model and the SVC firing angle model on various test systems. The SVC model for load flow analysis recommended by CIGRE and IEEE is the one widely used. This is the model that is used throughout this study. In this model, SVC characteristics within the limits are represented as a generator connected to a dummy bus that acts as a PV bus and is connected to high voltage bus through a reactor. For its operation outside the limits, this generator representation is, however, no longer valid. In such a case, changing reactance of SVC must be made constant.

Fig. 12.1 Basic structure of
SVC

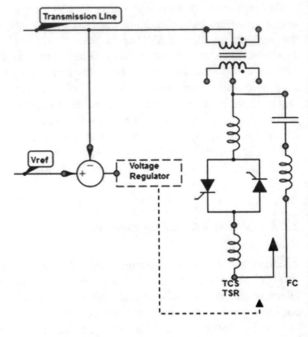

Fig. 12.2 SVC V-I
characteristics

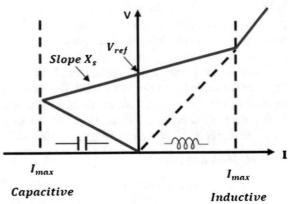

12.3 Problem Formulation

In order to design an SVC that could take care of the minor disturbances in the power system, all the possible contingencies are simulated and the contingencies are ranked according to a measure called Voltage Profile Index (VPI).

VPI is measured for each of the contingencies using the following equation:

$$VPI = \sum_{i=1}^{nbus} \left(\frac{V_i - V_{base}}{\Delta V_i^{max}} \right)^{2m}$$

Here V_i is the voltage at the ith bus after the disturbance has occurred. V_{base} is the voltage in the base case when there is no fault and the system is being operated in the steady state. ΔV_i^{max} is the maximum safe voltage deviation considered ideal by the system operators. It is taken as 0.2 p.u in this study. The exponent m takes the positive integers. This analysis takes the value of 'm' to be 2. In this study, the contingencies considered are outages of single lines. In the next step, a multicriteria function is developed, comprising of both operational objectives and investment costs. The worst three contingencies found out in the previous step are studied in detail to find out the location of SVC and its rating using Genetic Algorithm. The multiobjective function and the constraints described in the following lines are used in the formulation of Genetic Algorithm, which is described in the next section.

A. *Objective Function*

This consists of three objectives: two of them technical and one economical.

i. *Minimize total power losses in the network*

The line outage results in imbalance in reactive power. This leads to large flow of currents that in turn results in more losses in the transmission network. The location of SVC has to be such that all the reactive power injected into the grid minimizes power losses.

$$O_1 = P_{loss} = \sum_{k=1}^{nbr} g_k \left(V_i^2 + V_j^2 - 2V_i V_j \cos\left(\delta_i - \delta_j \right) \right)$$

ii. *Minimize total voltage deviations*

$$O_2 = VD = \sum_i^{nbus} \left| V_i - V_i^{ref} \right|$$

Here V_i^{ref} is taken as 1.0 p.u and $nbus$ is the total number of buses in the network.

iii. *Minimize investment costs*

$$O_3 = |Q_{SVC}|$$

where Q_{SVC} is the installed reactive power in p.u.

The investment cost of SVC is a function of SVC. Lower SVC rating results in lower investment costs.

B. *Constraints*

i. *Power Balance at each of the buses*

$$P_{G_i} - P_{L_i} = V_i \sum_{k=1}^{nbus} V_k (G_{ik} \cos(\theta_i - \theta_k) + B_{ik} \sin(\theta_i - \theta_k))$$
$$Q_{G_i} - Q_{L_i} = V_i \sum_{k=1}^{nbus} V_k (-B_{ik} \cos(\theta_i - \theta_k) + G_{ik} \sin(\theta_i - \theta_k))$$

where P_{G_i} and P_{L_i} denote the real power generation and the load at ith bus, respectively. Q_{G_i} and Q_i denote the reactive power generation and the load at ith bus, respectively. G_{ik} and B_{ik} are the real part and reactive part of the Y_{bus} matrix, respectively.

  ii. *Reactive Power Limits*: $Q_{min} \leq Q_{G_i} \leq Q_{max}$
 iii. *SVC reference value*

A positive value indicates that SVC generates reactive power and injects it into the grid while a negative sign indicates that SVC absorbs reactive power from the grid.

$$-Q_{Lmax} \leq Q_{G_i} \leq Q_{Cmax}$$

The SVC size is a variable that take any value in the above interval.

12.4   Problem Formulation Using Genetic Algorithm

The Genetic algorithm is a search method inspired by the biological evolution. It maintains a population of candidate solutions. Each individual is evaluated to get some measure of fitness and ranked according to the fitness function. At each generation, new individuals are formed by selecting more fit individuals based on certain selection criteria. Some of the population also undergoes changes to form new candidate solutions. Two commonly used genetic operations are crossover and mutation. Crossover is a mixing operator that gets its coordinates from selected parents. Mutation is used to search unexplored search space by randomly changing its position at one or more chromosomes. The rate at which crossover or mutation happens can be controlled [18].

A candidate solution (or chromosome) designed for the problem is a two component vector. The first component represents the location of SVC, which takes only integer values from 1 to number of buses in the network. The second component is the rating of SVC that takes values from -2 to 2 p.u.

Fitness Function: This is a measure to determine how close the candidate solutions are to the actual solution. In other words, it measures the quality of chromosomes. It is closely related to the objective function devised in the previous section. Since each of the objective functions have different units and different range of measures, we divide each of them by its base case value (system without SVC); assign weights to each of them and then add all of them together. For this problem, the overall fitness function is given by:

$$f(x, u) = 0.4 \frac{P_{loss}}{P_{loss, base}} + 0.4 \frac{VD}{VD_{base}} + 0.2 \times Q_{SVC}$$

where x denotes the variables to be determined and u represents all the dependent variables.

Note that in the event of contingency, protecting the system is more important rather than minimizing the cost of SVC. Hence, it is given less weight in the fitness function.

In order to solve this problem, the inbuilt MATLAB function to implement Genetic algorithm with default values of stopping criteria has been used. The general procedure followed for this implementation of Genetic algorithm can be summarized as follows [19]:

Start
Step 1: Read network data
Step 2: Solve Load flow using Newton Raphson method to get Power loss and voltage deviations in the base case
Step 3: Set genetic parameters and create initial population for SVC location and size.
Step 4: **While**(stopping condition is not reached):
        For each individual in current generation,
            calculate fitness function
        End (For)
        Selection(Current Generation, Population
size)
        Crossover (Selected Parents, Crossover Rate)
        Mutation (Current generation, Mutation rate)
        Increment Current Generation
    End (While)
Step 5: Show Solution
Stop

12.5 Case Study

The effectiveness of the proposed algorithm for optimally placing SVC devices has been tested on IEEE—9 bus and 30 bus systems using MATLAB R2014a. The line data and bus data for the system are taken from [20]. For each of the cases, the population size is taken as 50 and the number of generations is taken as 200.

A. *IEEE 9-Bus Test System*

The system comprises one slack bus, two generator buses, six load buses and nine lines. Each line outage is simulated and voltage profile index for each contingency is calculated and is tabulated (Table 12.1):

The case study is performed for the worst three contingencies which are determined by the VPI. Hence, the optimal SVC placement is implemented for the cases of loss of lines 9, 8 and 3.

The characteristics of GA convergence and the best individual generated in the final iteration for one of the cases (Loss of line no. 9) can be seen in figure (Fig. 12.3):

It was found out from the above simulation result that the installment of an SVC of 0.3836 p.u at bus 9 (Load bus index = 6 from the above simulation result) will mitigate the voltage deviations and ensure system security. This is evident from the following results of voltage profiles before and after the SVC installation at bus no. 9 (Fig. 12.4):

Similar analysis was done considering the outage of lines 8 and 3 (next worst contingencies) and it resulted in similar results. Overall analysis is provided in Table 12.2.

According to the results of Table 12.2, it can be seen that the power losses have been reduced to more than 50% in each of the cases. It is worth seeing noting that all the cases, except for that of loss of line 3 resulted in lesser voltage deviations with SVC installed. This is because its voltage profile was close to that in base case without SVC but it had huge losses. The SVC installed hence has reduced Power losses to large extent but has increased voltage deviations. These voltage deviations have also been such that the voltage security has been ensured, i.e., the voltages at all buses are in the permissible limits.

B. *IEEE 30-Bus Test System*

The system comprises one slack bus, 5 generator buses, 24 load buses and 41 lines. Each line outage is simulated and it is seen that Newton Raphson doesn't

Table 12.1 *VPI* for line outages in 9-bus system

Loss of line	VPI	Loss of line	VPI
9	0.19	6	0.0026
8	0.0243	5	0.0022
3	0.0163	2	0.0019
7	0.0043	–	–

Fig. 12.3 Working of GA
for 9 bus system (LO 9)

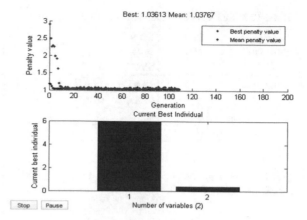

Fig. 12.4 Voltage profile
before and after SVC
installation at bus 9 (for LO 9)

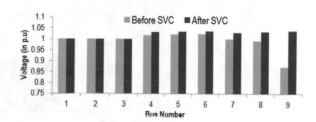

converge for four of the line outage simulations. For the rest of the line outages, they are illustrated in Table 12.3 in order of their VPI.

The first three contingencies have been considered to find out optimal location of SVC. Similar results were obtained for each of the cases. Hence, we show the graphical results of one of the contingencies, i.e., loss of line no. 36. For this larger system, the GA took comparatively larger time to converge and the convergence characteristics are shown in Fig. 12.5.

It has been found out that the SVC of 0.1043 p.u. rating at bus no. 30 (load bus index 24) would be the optimum allocation. It is evident from the following voltage profile results after this SVC installation is simulated.

Table 12.2 Summarized results of 9-bus system

Objective functions	Base case	LO 9		LO 8		LO 3	
		W/O SVC	With SVC	W/O SVC	With SVC	W/O SVC	With SVC
Total loss	4.5428	8.613	4.5469	13.0863	4.5549	9.3751	4.722
VD	0.0032	0.0176	0.0072	0.0061	0.0074	0.0034	0.0063
SVC location	0	9		9		5	
SVC rating	0	0.3836		0.4034		0.2719	

Table 12.3 Summarized results of IEEE 30-bus system

Objective functions	Base case	LO 9		LO 8		LO 3	
		W/O SVC	With SVC	W/O SVC	With SVC	W/O SVC	With SVC
Total loss	17.5985	19.8438	17.6452	18.0062	17.5663	77.7729	17.4856
VD	0.0396	0.0892	0.0549	0.0433	0.0415	0.0319	0.0432
SVC location	0	30		4		29	
SVC rating	0	0.1043		0.2368		0.022	

Fig. 12.5 Working of GA for 30 bus system (LO 36)

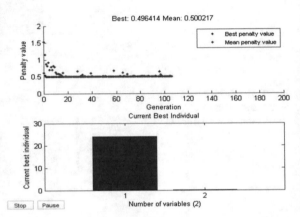

According to Fig. 12.6, it can be observed that the drop in voltages at buses 24–30 have been raised after SVC has been installed. All the three line outages are simulated and it has been seen that all of them resulted in voltage security. Overall analysis is illustrated in Table 12.4. This table represents that the total losses have been reduced in all the cases. It has also been noticed that the voltage deviations also decreased in first two cases but has increased to some extent in the case 3 (LO 37). This might be because the Power losses have been quite large before SVC and the installation of SVC has reduced the total power losses by four times, which resulted in a slight increase in voltage deviations.

Fig. 12.6 Voltage profile before and after SVC installation at bus 30 (for LO 36)

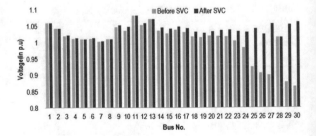

Table 12.4 VPI for line outages in 30-bus test system

Loss of line	VPI	Loss of line	VPI
36	0.6169	25	0.0042
1	0.0565	9	0.0035
37	0.0074	15	0.0022
28	0.006	–	–

12.6   Conclusion

In this paper, we proposed a method for allocation of the SVCs of different ratings and their optimal locations. Our method optimizes Power losses and voltage deviations for small and medium size systems using genetic algorithm. Analysis of voltage profile is also provided for different contingencies to notice the effect of proposed algorithm in maintaining the voltage magnitude.

We implemented our method on IEEE 9-bus and 30-bus test systems. After installing these SVCs on their corresponding optimal location, the power losses and voltage deviations have been compared thoroughly. The simulation results verifies that such placement of SVC resulted in optimum power losses and voltage deviations which were considerably close to the globally optimum solution obtained in the base case. For the analyzed test systems, one or more locations of SVCs also have been simulated but it has been found out that for the systems of this scale, one SVC would suffice to attain the goal of voltage security of the system before and after one critical line outage. It has been found out that the obtained SVCs location and rating maintained voltage security and minimized power loss in the event of contingency. Although the proposed method has been tested on only two test systems, the satisfactory results in both the systems indicate that this approach can be generalized for other test systems.

References

1. A.K. Mohanty, A.K. Barik, Power system stability improvement using FACTS devices. Int. J. Modern Eng. Res. **1**(2), 666–672 (2011)
2. M. Cvetkovic, M.D. Ilić, Energy based transient stabilization using FACTS in systems with wind power, in *Proceedings of IEEE Power and Energy Society General Meeting* (2012)
3. K.D. Bachvochin, M.D. Ilić, Automated passivity based control law derivation for electrical euler lagrange systems and demonstration on three phase AC/ DC converter, in *Technical Paper* (2014)
4. A. Sargolzaei, K.K. Yen, M.N. Abdelghani, Preventing time-delay switch attack on load frequency control in distributed power systems. IEEE Trans. Smart Grid **7**(2), 1176–1185 (2016)
5. S. Noei et al., A decision support system for improving resiliency of cooperative adaptive cruise control systems. Procedia Comput. Sci. **95**, 489–496 (2016)
6. A. Abbaspour et al., Detection of fault data injection attack on UAV using adaptive neural network. Procedia Comput. Sci. **95**, 193–200 (2016)

7. M.H. Amini, M. Hadi et al., Distributed security constrained economic dispatch, in *IEEE Innovative Smart Grid Technologies-Asia (ISGT ASIA)* (IEEE, 2015)
8. A. Mohammadi, M. Mehrtash, A. Kargarian, Diagonal quadratic approximation for decentralized collaborative TSO+DSO optimal power flow. IEEE Trans. Smart Grid (2018). https://doi.org/10.1109/TSG.2018.2796034
9. M.H. Amini et al., Application of cloud computing in power routing for clusters of microgrids using oblivious network routing algorithm, in *19th European Conference on Power Electronics and Applications (EPE'17 ECCE Europe)* (2017)
10. M.H. Amini et al., Load management using multi-agent systems in smart distribution network. IEEE Power Energy Soc. General Meeting (2013)
11. S. Bahrami, M.H. Amini, M. Shafie-khah, J.P.S. Catalao, A decentralized electricity market scheme enabling demand response deployment. IEEE Trans. Power Syst. (2017). https://doi.org/10.1109/TPWRS.2017.2771279
12. M.H. Amini, K.G. Booojeni, T. Dragicevic, A. Nejadpak, S.S. Iyengar, F. Blaabjerg, A comprehensive cloud-based real-time simulation framework for oblivious power routing in clusters of DC Microgrids, in *2nd IEEE International Conference on DC Microgrids (ICDCM)* (2017)
13. A. Sode-Yome, Nadarajah Mithulananthan, Comparison of shunt capacitor, SVC and STATCOM in static voltage stability margin enhancement. Int. J. Electr. Eng. Educ. **41**(2), 158–171 (2004)
14. H. Ambriz-Perez et al., Advanced SVC models for Newton-Raphson load flow and Newton optimal power flow studies. IEEE Trans. Power Syst. **15**(1), 129–136 (2000)
15. R. Sirjani, A. Mohamed, H. Shareef, Optimal allocation of shunt Var compensators in power systems using a novel global harmony search algorithm. Int. J. Electr. Pow. Energy Syst. **43**, 562–572 (2012)
16. R. Selvarasu et al., SVC placement for voltage constrained loss minimization using self-adaptive Firefly algorithm. Arch. Electr. Eng. **62**(4), 649–661 (2013)
17. J. Kueck et al., Reactive power from distributed energy. Electr. J. **19**(10), 27–38 (2006)
18. S. Dixit, L. Srivastava, G. Agnihotri, Minimization of power loss and voltage deviation by SVC placement using GA. Int. J. Control Autom. **7**(6), 96–108 (2014)
19. I. Pisica, C. Bulac, L. Toma, M. Eremia, Optimal SVC placement in electric power systems using a genetic algorithms based method, in *Proceedings International Conference on Power Technologies* July (2009)
20. H. Saadat. *Power System Analysis* (WCB/McGraw-Hill, 1999)

Chapter 13
Decentralized Control of DR Using a Multi-agent Method

Soroush Najafi, Saber Talari, Amin Shokri Gazafroudi,
Miadreza Shafie-khah, Juan Manuel Corchado and João P. S. Catalão

13.1 Introduction

Today, the environmental concerns, and the economic aspects of using conventional energy resources cause governments, technology providers, and academic communities to use more renewable energy resources. These resources do not have remarkable pollution and operating costs. Basically, the hourly demand response (DR) program in electricity markets could provide significant benefits to market participants and customers. Such benefits include lower hourly market prices, lower

S. Najafi
Department of Electrical Engineering, Isfahan University of Technology, Isfahan, Iran
e-mail: soroush.najafy88@gmail.com

S. Talari · M. Shafie-khah (✉) · J. P. S. Catalão
C-MAST, University of Beira Interior, 6201-001 Covilhã, Portugal
e-mail: miadreza@ubi.pt; miadreza@gmail.com

S. Talari
e-mail: saber.talari@ubi.pt

J. P. S. Catalão
e-mail: catalao@fe.up.pt

A. S. Gazafroudi · J. M. Corchado
BISITE Research Group, University of Salamanca, Edificio I+D+i, 37008 Salamanca, Spain
e-mail: shokri@usal.es

J. M. Corchado
e-mail: corchado@usal.es

J. P. S. Catalão
INESC TEC and the Faculty of Engineering of the University of Porto, 4200-465 Porto, Portugal

J. P. S. Catalão
INESC-ID, Instituto Superior Técnico, University of Lisbon, 1049-001 Lisbon, Portugal

© Springer International Publishing AG, part of Springer Nature 2018
M. H. Amini et al. (eds.), *Sustainable Interdependent Networks*,
Studies in Systems, Decision and Control 145,
https://doi.org/10.1007/978-3-319-74412-4_13

volatility in hourly market prices, enhanced system reliability and a smaller chance for the market power exertion by generating companies (GENCO), as a purchaser plays more active role in power system operations.

DR is a program that guarantees customers maximize their benefit with participating in the electrical market as an increase power system efficiency with integrating renewable generation. However, according to [1], few DR programs are developed for residential customers. Nonetheless, these programs are related to smart grids technologies, such as smart metering. Fortunately, a great progress occurs in recent years [2]. For convincing customers to apply this program and change the time for buying purchasing energy from the electrical market, paying the price in off-peak hours is lower than peak hour's price.

Several DR strategies for residential and non-residential customers are presented in [3–6]. They solve peak load shaving problem. In [7], other proposed strategies in the USA for commercial and industrial customers have been reviewed. They are direct load control, real-time pricing, and time-of-use programs. In [8], the optimal performance of the Household Energy Management (HEM) system is investigated, and the household's participation in both incentive-based and price-based DR is modeled. In [9], the benefits of a market-based DR program in reliability for the future of China and the improvement of the existing DR program in the USA are presented. A real-time price (RTP)-based automated demand response strategy for a PV-assisted electric vehicle charging station is proposed at [10]. The interaction between utility companies and multiple customers in the smart grid by modeling the DR program is presented at [11].

The main idea of this chapter is a decentralized scheme with considering DR programs. Each agent is participating in the electrical market with implementing a market strategy. In this strategy, DR-enabled customers will participate in the market based on their energy requirements. The market will sell energy to agents with considering their bidding, similar to the model presented in [12] for plug-in electric vehicles. This strategy will minimize the cost of buying energy for each agent with covering consumer's consumption. Two different strategies, the centralized and the decentralized approaches, are mentioned in [12]. An agent-based fully distributed DC Security Constrained Optimal Power Flow (DC-SCOPF) approach is presented in [13]. In [14], a decentralized energy trading algorithm is proposed with proper control signals using the Lagrange relaxation technique for maximum entities profit. In [15], a self-decision-making method for load management (LM) which utilized the multi-agent systems considering the upstream grid, distributed generation (DG) and demand response resources (DRR) is presented.

In this chapter, first, we introduce a centralized and decentralized approach to control the consumption and state their advantages and disadvantages. In Sect. 13.3, DR program will be defined. The Q-learning algorithm presented in Sects. 13.4 and 13.5, case studies and various scenarios are introduced. Finally, in Sect. 13.6, the conclusion of whole book chapter is presented.

13.2   Centralized and Decentralized Approach

There are several strategies for cost minimization that consumers can adopt. In [12], two different strategies are introduced, centralized and decentralized approaches. These approaches are similar in the term of energy cost and achieved valuable results.

In the central model, there is a central agent, e.g., an aggregator, whose objective is controlling all other energy consumption agents and taking care of serving the electricity demand while trying to minimize the cost. The centralized approach is based on bidirectional communication between agents and aggregator agent. For participating in the electrical market, the customers that enabled their DR program send their information to the central agent. The central agent scheduled energy consumption and sent a control signal to the user. Although this approach motivates us to optimal cost but with increasing customer number, this method can be less valuable due to difficult information processing and high computational calculation [12]. In Fig. 13.1, the relationship between DR agents and electrical market in centralized method is presented.

On the other hand, the decentralized approaches required lower communication, just being enough to transfer the price in order to control energy consumption, as presented in [16–19] for electric vehicles. However, unacceptable outputs can happen, like the possibility of simultaneous reactions, such as avalanche effects or errors in forecasting the consumer's attitude in relation to the price signals. This occurs due to the fact that the impact of the DR-enabled customer's bids is not taken into account. Hence, this type of strategy is probably only suitable for low penetrations.

To avoid the mentioned unacceptable results, bidirectional communication will be used in this approach. Although this approach is rising communication requirements, it will be solved load synchronization problems with the bidding process, as it mentioned in [20], or with an iterative process such as in [21, 22]. Each agent determines its bids according to their energy requirements, and the

Fig. 13.1 Relationship between DR agents and electrical market in centralized method

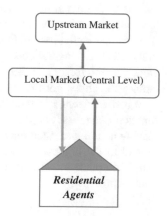

Table 13.1 Comparing advantages and disadvantages of different control method

Control method	Advantages	Disadvantages
Centralized, bidirectional communication	1. Achieve optimal outcomes	1. Scalability and complexity for the aggregator
	2. Uncertainty for different consumers can be better managed	2. High communication requirements
		3. User privacy
Decentralized, unidirectional communication	1. Low communication requirements	1. Undesirable outcomes, hence impact of consumers demand on prices is neglected
	2. Properties of battery modeled in a more detailed way	2. Only effective for low agents penetration
Decentralized, bidirectional communication	1. Work close to real-time and agents demands is take account	High communication requirements
	2. User privacy is guaranteed	

electrical market answers to the bids by communicating the clearing price. In [23], a new decentralized method is introduced based on ISO and consumers. With this method, the privacy of all consumers is guaranteed. In Table 13.1, advantages and disadvantages of different control methods are presented.

13.3 Q-Learning

Reinforcement learning dates back to the early days of cybernetic and works in the statistic. In the last decade, it has attracted rapidly increasing in the machine learning and artificial intelligence communities. Reinforcement learning is the problem faced by an agent that must learn behavior through trial-and-error interactions with a dynamic environment [24]. Q-learning is a model-free reinforcement learning, and it is typically easier to implement.

Each residential load defined as an agent. Agents should learn how to participate in the electrical market and optimize their cost, simultaneously. To this end, in this work, a reinforcement learning algorithm so-called Q-learning algorithm is employed. This algorithm has been firstly developed by Watkins [25].

When an agent i is modeled by a Q-learning algorithm, it keeps in memory a function $Q_i : A_i \to R$ such that $Q_i(a_i)$ represents it will obtain the expected reward if playing action a_i. It then plays with a high probability the action it believes is going to lead to the highest reward, observes the reward it obtains and uses this observation to update its estimate of Q_i. Suppose that the tth time the game is played, the joint action $(a_1^t, \ldots, a_n^t)$ represents the actions the different agents have taken.

Fig. 13.2 Simulation of reinforcement learning agents interacting with a matrix game [26]

1] Determine $t = 0$.
2] Begin $Q_i(a_i) = 0$ ∀$i \in \{1, \cdots, n\}$ and ∀$a_i \in A_i$.
3] $t \leftarrow t + 1$.
4] Choose for each factor i an action a'_i by applying a ε-Greedy policy.
5] Play the game with the common actions $(a'_1, \cdots, a'_n)$.
6] Look at each factor i the reward $r_i(a'_1, \cdots, a'_n)$ it has taken.
7] Update for each factor i its Qi-function according to:
$\quad Qi(a'_i) \leftarrow Qi(a'_i) + \alpha ti(ri(a'_1, \cdots, a'_n) - Qi(a'_i))$
8] If enough number of games has been played, then stop. Otherwise, back to step 3.

After the game is played and the different rewards r_i have been observed, agent i updates its Q_i-function according to Eq. 1:

$$Q_i\left(a_i^t\right) \leftarrow Q_i\left(a_i^t\right) + \lambda_i^t\left(r_i\left(a_1^t, \ldots, a_n^t\right) - Q_i\left(a_i^t\right)\right) \qquad (13.1)$$

$\lambda_i^t \in [0, 1]$ is the correction degree. If $\lambda_i^t = 1$, the agent supposes that the expected reward it will get by taking action $a_i = a_i^t$ in the next game is equal to the reward it just observed. If $\lambda_i^t = 0$, it means the agent does not use its last observation to update the value of its Qi-function. Regarding the fact that the Q-learning algorithm provides the agent information to know the most profitable actions, a policy for decision making is required. This is justified by the fact that there is no decision rule associated with the process on beforehand. By using a Greedy policy, agents will exploit the available information, always choosing the actions that maximize the reward.

Figure 13.2 shows a tabular version of the algorithm that simulates reinforcement learning driven agents interacting with a matrix game. The number of games after which the simulation should be stopped (step 8 of the algorithm) depends on the purpose of the study.

13.4 Agents Actions

For the DR agents to adapt their bids to the price in the market, a set of actions must be taken in order to measure the sensitivity of their energy requirements and adapt it to a price. In [27], a pricing function in which the cost of purchasing energy is dependent on the level of consumption is presented. The price function takes into account the effect of the customers' consumption level, so the customers will have a multiple price rating according to their consumption level. Thus, the value of p_{state} is adapted considering an independent constant variable, and another constant multiplied to p_{state}. The adapted formulation is presented below.

$$p_{state} = B_1 + B_2 * \left(\frac{t}{T} * \frac{P_C^t + P_{TS}^t}{P_L} \right) \quad\quad (13.2)$$

In this scenario, B_1 represents the agents' interest in buying energy, i.e., the price that agents consider to be low enough to purchase energy even if there is no urge to buy it. B_2 represents the sensitivity to the urge of buying energy of the willingness to buy electricity. Meanwhile, the state variables are energy-requirement dependent; actions may have a wider range of values previously defined.

13.5 Case Study

In this book chapter, three houses that want to participate in DR programs have been assumed. As you can see in Fig. 13.3, agents send their bids to electrical market, and after determining market clearing price, it announced to agents.

Since the scheme of this work is directed to residential loads, those household loads will be classified into two categories: controllable loads and critical loads. Critical loads are the loads that cannot be controlled, i.e., loads that must mandatorily be served at every time-step. Otherwise, major impacts will be observed in the consumer's lifestyle. Loads such as lighting and refrigeration are part of this type of loads.

On the other side, controllable loads consist of loads that can be controlled without major impacts on the end-user's lifestyle. Loads such as HVAC systems, clothes dryer, or water heater can be served at several different times without causing much impact on the end-user's lifestyle. Thus, this type of loads can be seen as loads shiftable in time.

Each house buys its critical load at the market clearing price, and after that, they participate in the market and buy energy for the shiftable load based on their bids.

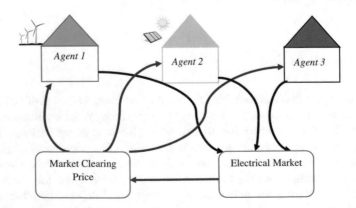

Fig. 13.3 Relationship between DR agents and electrical market in decentralized method

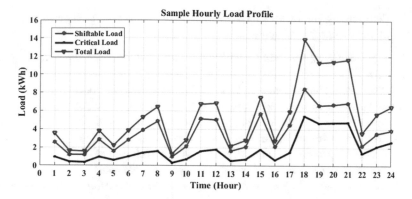

Fig. 13.4 Hourly load profile

They can be a DR agent. An example of load profile for each agent is presented in Fig. 13.4. We always consider critical load is less than %50 of the total load. Agent 1 and agent 2 can have small renewable energy generation in their house. The agents with renewable energy did not participate in the electricity market, and we model their generation as a negative load. Moreover, because of the day-ahead market, we did not consider uncertainty for these RES. Even if we consider that, it does not have a major effect on our study. Solar and wind generation is shown in Fig. 13.5. For the sake of simplicity, it is assumed that total generation for wind and solar is constant for all of the iterations and hence for all day in a year.

As it stated in Sect. 13.4, for participating in the electrical market, Q-learning algorithm should determine 2 parameters (B_1 and B_2). Each pair of these parameters can account for an action. In this book chapter, there are two possible B_1 (36, 38) and three possible B_2 (1, 10, 50). Market clearing price is not known for agents. They participated in the market and, based on their bidding, can buy energy from the network.

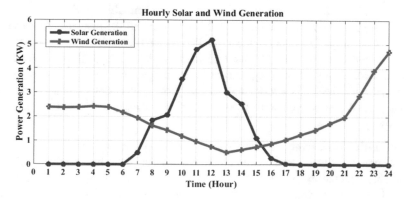

Fig. 13.5 Hourly solar and wind generation

For determining the effect of DR program and renewable energy in total cost, eight different scenarios are proposed:

- *1st Scenario–without wind, without solar, without variable load (Base Scenario)*
- *2nd Scenario–without wind, without solar, with variable load*
- *3rd Scenario–with wind, without solar, without variable load*
- *4th Scenario–with wind, without solar, with variable load*
- *5th Scenario–without wind, with solar, without variable load*
- *6th Scenario–without wind, with solar, with variable load*
- *7th Scenario–with wind, with solar, without variable load*
- *8th Scenario–with wind, with solar, with variable load (Comprehensive Scenario)*

13.6 Results and Discussion

In this section, simulation result for all scenarios would be presented. In scenarios without variable load, since we do not have a lot of agents and a lot of states, algorithm converged very soon (during 20 iterations). Although, if we have a real condition with variable load, we always have some offset and algorithm cannot converge correctly. Nevertheless, the agents' cost or average cost between scenarios with variable load and scenarios with constant load has not a meaningful difference.

1st Scenario: without wind, without solar, without variable load (Base Scenario)
In this scenario, agents do not have wind or solar generation. For comparison with other scenarios, that load is assumed to be constant during all iterations. As you can see in Fig. 13.6, after the transition in first 10 iterations, the algorithm has converged.

2nd Scenario: without wind, without solar, with variable load
Variable load is the difference between this scenario and the first one. As expected, it did not converge because of this volatility. Moreover, since any agents did not have a generation, the cost for each of them is the same. Although we have a variable load, average cost or agent cost for each agent does not have a significant difference with the 1st scenario, and it can prove algorithm in real condition is working good. Agent cost and average cost are shown in Figs. 13.7 and 13.8 respectively.

3rd Scenario: with wind, without solar, without variable load
This scenario is similar to 1st scenario except that agent 3 has small wind generation. This generation can reduce average cost and cost for agent 3. Figures 13.9 and 13.10 represent average cost and agent cost, respectively. As we expect, agent 3

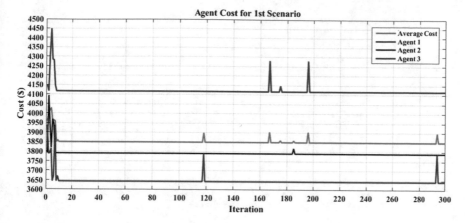

Fig. 13.6 Average and agent cost for 1st scenario

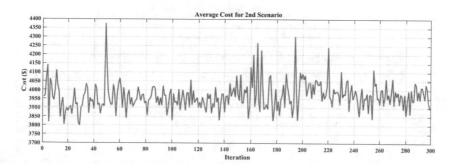

Fig. 13.7 Average cost for 2nd scenario

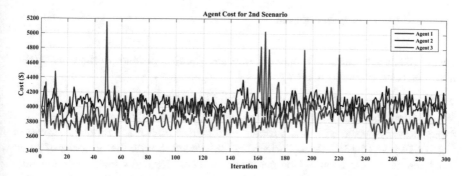

Fig. 13.8 Agent cost for 2nd scenario

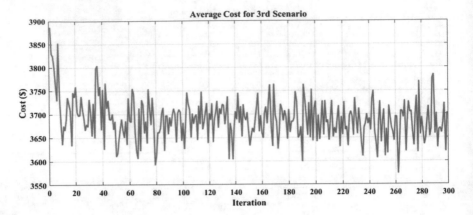

Fig. 13.9 Average cost for 3rd scenario

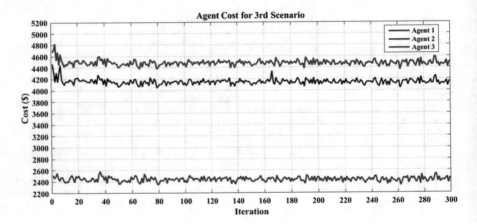

Fig. 13.10 Agent cost for 3rd scenario

cost is much less than other agents. Because of constant load profile, low fluctuations are observable in these figures.

4th Scenario: with wind, without solar, with variable load

Wind generation and variable load are considered in this scenario. Wind generation causes lower average cost, and variable load causes high fluctuations as shown in Fig. 13.11 and 13.12. This scenario is the 2nd scenario with the presence of small wind generation for agent 3. Comparing agent 3 cost for 2nd scenario and this one, cost reduction can be seen. Because of load randomly behavior, the average cost is not as low as the 3rd scenario, but cost reduction during first 20 iterations is obvious. In both 3rd and 4th scenarios, agent 3 has a lower cost because of wind generation as expected.

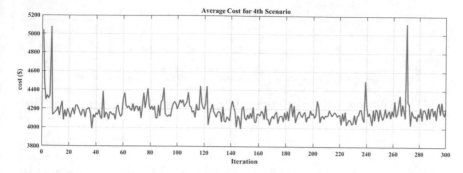

Fig. 13.11 Average cost for 4th scenario

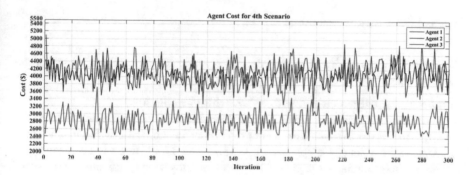

Fig. 13.12 Agent cost for 4th scenario

5th Scenario: without wind, with solar, without variable load

In this scenario, agent 2 has small photovoltaic generation as shown in Fig. 13.4. Also, load is assumed to be constant during iterations. Average cost and agent cost are shown in Figs. 13.13 and 13.14. In this scenario, agent 2 reduces its cost

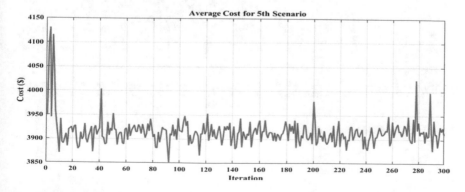

Fig. 13.13 Average cost for 5th scenario

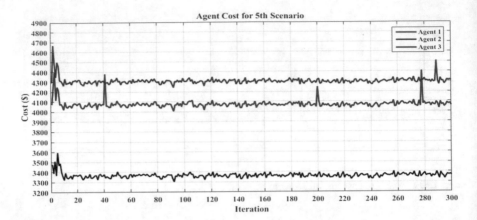

Fig. 13.14 Agent cost for 5th scenario

because of PV generation. Same as the 3rd scenario, the fluctuation is low for the constant load.

6th Scenario: without wind, with solar, with variable load
Same as previous scenario, agent 2 has PV generation. This generation reduces the average and agent 2 cost. This scenario is more in the real condition with considering variable load. Comparing Figs. 13.15 and 13.13, there is no meaningful difference between the average cost for two scenarios. Hence, the algorithm works

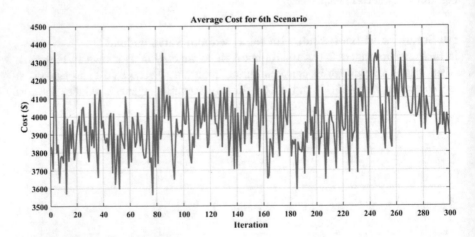

Fig. 13.15 Average cost for 6th scenario

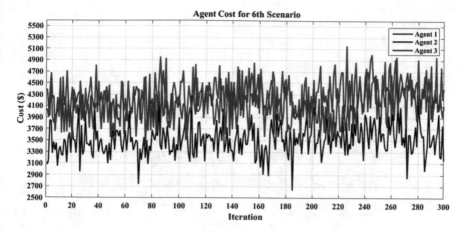

Fig. 13.16 Agent cost for 6th scenario

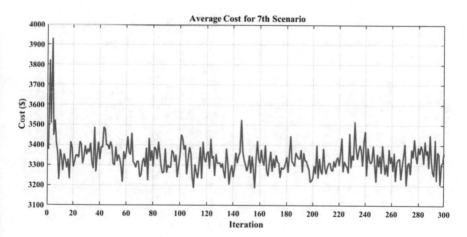

Fig. 13.17 Average cost for 7th scenario

in real market condition as good as a constant load. Variable load causes high fluctuations as shown in Fig. 13.15. Agent cost is presented in Fig. 13.16.

7th Scenario: with wind, with solar, without variable load

In this scenario, agent 2 has PV generation and agent 3 has wind generation. Average cost is lower than previous scenarios as can be seen in Fig. 13.17. Because of considering constant load profile during all iterations, average and agent cost have low fluctuation (Fig. 13.18).

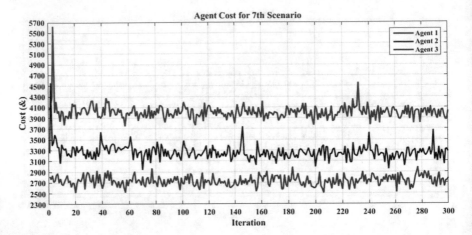

Fig. 13.18 Agent cost for 7th scenario

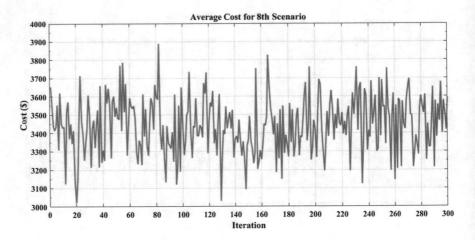

Fig. 13.19 Average cost for 8th scenario

8th Scenario: with wind, with solar, with variable load (Comprehensive Scenario)

The comprehensive scenario consists of all of the residential generation (PV and wind generation) and variable load profile for each iteration. Average cost and agent cost are presented in Figs. 13.19 and 13.20, respectively. As we said before, variable load causes high fluctuation in average and agent cost.

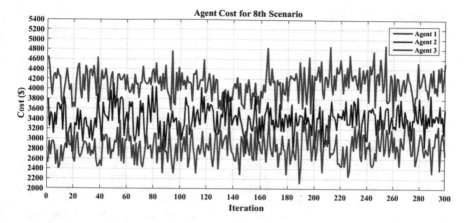

Fig. 13.20   Agent Cost for 8th Scenario

13.7   Conclusion

In this chapter, decentralized control of DR using a multi-agent method is presented. Three residential loads and 2 residential generations were considered. Without losing the integrity, we can increase the number of residential load and generation. The residential load is modeled with 2 parts, critical load and controllable load. With implementing DR program, the controllable load can shift to the time with the lower price. For determining bidding strategy for buying power from the network, we applied a reinforcement learning (Q-learning) algorithm. This algorithm can guaranty end-user privacy and independent them from aggregator company. If load profile changes for each iteration (real condition) average, agent cost does not have so much difference. It means with applying the Q-learning algorithm, we can reach minimum cost for buying energy from the network. Eight different scenarios are investigated in this chapter. If each residential load or each agent applying small renewable DG, their cost reduced significantly. In the comprehensive scenario, 2 agents have small DG. Hence, the cost reduction is more than other scenarios.

Acknowledgements The work of Saber Talari, Miadreza Shafie-khah and João P.S. Catalão was supported by FEDER funds through COMPETE 2020 and by Portuguese funds through FCT, under Projects SAICT-PAC/0004/2015—POCI-01-0145-FEDER-016434, POCI-01-0145-FEDER-006961, UID/EEA/50014/2013, UID/CEC/50021/2013, and UID/EMS/00151/2013. Also, the research leading to these results has received funding from the EU Seventh Framework Programme FP7/2007-2013 under grant agreement no. 309048.

Amin Shokri Gazafroudi and Juan Manuel Corchado acknowledge the support by the European Commission H2020 MSCA-RISE-2014: Marie Sklodowska-Curie project DREAM-GO Enabling Demand Response for short and real-time Efficient And Market Based Smart Grid Operation—An intelligent and real-time simulation approach ref. 641794. Moreover, Amin Shokri Gazafroudi acknowledge the support by the Ministry of Education of the Junta de Castilla y León and the European Social Fund through a grant from predoctoral recruitment of research personnel associated with the research project "Arquitectura multiagente para la gestión eficaz de redes de energía a través del uso de técnicas de intelligencia artificial" of the University of Salamanca.

References

1. Assessment of demand response and advanced metering. Washington, DC, USA, Tech. Rep., Dec. 2012
2. Y. Li, B.L. Ng, M. Trayer, L. Liu, Automated residential demand response: Algorithmic implications of pricing models. IEEE Trans. Smart Grid 3(4), 1712–1721 (2012)
3. H. Aalami,M.P. Moghadam, G R. Yousefi, Optimum time of use program proposal for Iranian power systems, in *Proceedings International Conference on Electrical Power Energy Conversion Systems*, November 2009
4. S. Ashok, R. Banerjee, Optimal operation of industrial cogeneration for load management. IEEE Transactions Power System 18(2), 931–937 (2003)
5. J. Joo,S. Ahn, Y. Yoon,J. Choi, Option valuation applied to implementing demand response via critical peak pricing, in *Proceedings IEEE Power Energy Society General Meeting*, Jun 2007
6. A.B. Philpott, E. Pettersen, Optimizing demand-side bids in day-ahead electricity market. IEEE Trans. Power Syst. 21(2), 488–498 (2006)
7. https://www.ferc.gov/legal/staff-reports/2010-dr-report.pdf
8. M. Shafie-khah et al., Optimal behavior of responsive residential demand considering hybrid phase change materials. Appl. Energy 163, 81–92 (2016)
9. F. Wang et al., The values of market-based demand response on improving power system reliability under extreme circumstances. Appl Energy 193 220–231 (2017)
10. Q. Chen et al., Dynamic Price vector formation model based automatic demand response strategy for PV-assisted EV charging station. *IEEE Trans. Smart Grid* (2017)
11. F. Kamyab et al., Demand response program in smart grid using supply function bidding mechanism. IEEE Trans. Smart Grid 7(3), 1277–1284 (2016)
12. M.G. Vayá, L B. Roselló, G. Andersson, Optimal bidding of plug-in electric vehicles in a market-based control setup, in *Power Systems Computation Conference* (2014)
13. J. Mohammadi, G. Hug, S. Kar, Agent-based distributed security constrained optimal power flow, in *IEEE Transactions on Smart Grid* (2016)
14. S. Bahrami, M.H. Amini, A decentralized framework for real-time energy trading in distribution networks with load and generation uncertainty (2017), arXiv:1705.02575
15. M.H. Amini, B. Nabi, M.-R. Haghifam, Load management using multi-agent systems in smart distribution network, in *IEEE Power and Energy Society General Meeting (PES)* (IEEE, 2013)
16. M.G. Vayá, Roselló, G. Andersson, Centralized and decentralized approaches to smart charging of plug-in vehicles, in *IEEE Power and Energy Society General Meeting* (2012)
17. N. Rotering, M. Ilic, Optimal charge control of plug-in hybrid electric vehicles in deregulated electricity markets. IEEE Trans. Power Syst. 26(3), 1021–1029 (2011)
18. S. Bashash, S.J. Moura, J.C. Forman, H.K. Fathy, Plug-in hybrid electric vehicle charge pattern optimization for energy cost and battery longevity. J. Power Sources 196(1), 541–549 (2011)
19. A. Hoke, A. Brissette, D. Maksimovic, A. Pratt, K. Smith, Electric vehicle charge optimization including effects of lithium-ion battery degradation, in *IEEE vehicle power and propulsion conference (2011)*
20. J.K. Kok, M.J.J. Scheepers, I.G. Kamphuis, *Intelligence in electricity networks for embedding renewables and distributed generations, Intelligent Infrastructures, Intelligent Systems, Control and Automation: Science and Engineering*, vol. 42 (Springer, Netherlands, 2010), pp. 179–209

21. Z. Ma, D.S. Callaway, I.A. Hiskens, Decentralized charging control of large populations of plug-in electric vehicles. IEEE Trans. Control Syst. Technol. **21**(1), 67–78 (2013)
22. L. Gan, U. Topcu, S. Low, Optimal decentralized protocol for electric vehicle charging. IEEE Trans. Power Syst. **28**(2), 940–951 (2013)
23. S. Bahrami, M.H. Amini, M. Shafie-khah, J.P.S. Catalao, A decentralized electricity market scheme enabling demand response deployment. IEEE Trans. Power Syst. (2017). https://doi.org/10.1109/TPWRS.2017.2771279
24. L.P. Kaelbling, M.L. Littman, A.W. Moore, Reinforcement learning: a survey. J. Artif. Int. Res. **4**, 237–285 (1996)
25. C.J.C.H. Watkins, Learning from delayed rewards. Ph.D. thesis, King's College, Cambridge, 1989
26. T. Krause, et al., A comparison of Nash equilibria analysis and agent-based modelling for power markets. Int. J. Electr. Power Energy Syst. **28**(9), 599–607 (2006)
27. H.T. Haider, O.H. See, W. Elmenreich, Residential demand response scheme based on adaptive consumption level pricing. Energy **113**, 301–308 (2016)

Chapter 14
Complex Distribution Networks: Case Study Galapagos Islands

Diego X. Morales, Yvon Besanger and Ricardo D. Medina

Nomenclature

η_B	Battery performance
ADA	Advanced distribution automation
ADMS	Advanced distribution management system
ARCONEL	Ecuadorian national electricity regulator
BESS	Battery energy storage system
BMS	Battery management system
C_{ch}	Cost of charging batteries
C_{dch}	Savings originated for discharging batteries
C_{en}	Energy cost
$C_{load\ ppv}$	Cost of supply a load in the presence of photovoltaic generation
Co	Capital cost, including purchase, installation, and maintenance
C_T	Total cost of the system
D(i)	Typical number of days the ith season
DR	Demand response
DSM	Demand side management
E_{Lmed}	Absorbed energy during the medium consumption period
E_{Loff}	Absorbed energy during the off-peak period
E_{Lpeak}	Absorbed energy during the peak period
EM	Electrical motorbike
EV	Electric vehicle
G	Population growing
IC	Induction cooker

D. X. Morales (✉)
Smart Grid Research Group, Universidad Católica de Cuenca and Univ.
Grenoble Alpes, CNRS, Grenoble INP, F-38000 Grenoble, France
e-mail: Diego.Morales-Jadan@g2elab.grenoble-inp.fr

Y. Besanger
G2Elab, Univ. Grenoble Alpes, CNRS, Grenoble INP, F-38000 Grenoble, France

R. D. Medina
Instituto de Energía Eléctrica – Universidad Nacional de San Juan, San Juan, Argentina

© Springer International Publishing AG, part of Springer Nature 2018
M. H. Amini et al. (eds.), *Sustainable Interdependent Networks*,
Studies in Systems, Decision and Control 145,
https://doi.org/10.1007/978-3-319-74412-4_14

Kcrit	Maximum value for unit cost for the admitted capacity loss
MV	Medium voltage
N	Number of years of the studied period
P_{BAT}	Power supplied for battery systems
P_{EGRID}	Demanded power from the grid
P_{LOAD}	Demanded power for load
Pn	Rated power in KVA
P_{PV}	Power supplied for PV sources
PQ	Load fed with active and reactive power curves
Pri_{med}	Energy price of medium consumption period
Pri_{off}	Energy price of off-peak period
Pri_{peak}	Energy price of peak period
$Pri_{ppv\ med}$	Photovoltaic energy consumption during the medium consumption period
$Pri_{ppv\ off}$	Photovoltaic energy consumption during the off-peak period
$Pri_{ppv\ peak}$	Photovoltaic energy consumption during the peak period
PV	Photovoltaic sources
SG	Smart grid
SOC	State of charge
TOU	Time of use
Wp	Watts peak
Z	Ageing factor
β	Discount rate
η_b	Battery performance

14.1 Introduction

The Galapagos Islands are an archipelago of volcanic islands distributed on either side of the equator line in the Pacific Ocean, located at 926 km west of Ecuador. The Galapagos Islands and their surrounding waters are an Ecuadorian Province, as well as a national park and a biological marine reserve. Since 1978, they are declared as World Heritage [1] (Fig. 14.1).

Due to the growth of the population, there are several social, economic, and environmental problems, which endanger the environment conservation of the islands. According to the Ecuadorian National Institute Statistic and Census on 2010 in the Galapagos Province, there were 25.274 people, of which 15.393 live at Santa Cruz, 7.475 at San Cristobal at Isabela 2.256, and 150 at Floreana Island [3].

The Ecuadorian government wants to preserve its ecological heritage, hence with the participation of stakeholders such as the Ministry of Energy and Renewable Energy and Galapagos Government Council; some initiatives such as replacing the conventional vehicles by electrical ones and the gas stoves by induction ones

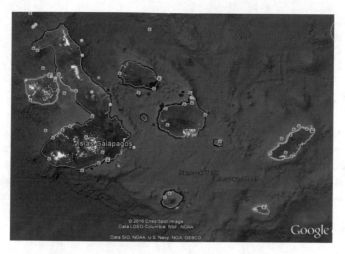

Fig. 14.1 Galapagos satellite view [2]

have been released. In addition, strong integration of distributed generation, deployment of an Advanced Distribution Management System (ADMS), and a future deployment of smart meters were considered to improve the general services provided to the island's population. Presently, one of these initiatives is the EcoSmart Project [4], which is intended to improve the preservation of the Galapagos Islands and to reduce the environmental footprint, becoming a world-class reference in the management of the energy and sustainability. In order to accelerate the islands through efficiency and quickly transition toward a smart grid (SG), some Advanced Distribution Automation (ADA) functions such as self-healing [5], demand-side management (DSM) [6], Volt–Var Control [7], and Network Reconfiguration [8] have to be implemented.

In [9–11], a smart grid is defined as "An electricity network that can intelligently integrate the actions of all actors connected to it-generators, consumers and those that do both functions to deliver sustainable, economical and safe power supply efficiently."

Worldwide improvements achieved in the Information and communications technology field have enabled the implementation of smart grid, which are significant tools recommended for the traditional electric power systems. For example, Denmark, a pioneer in the development of smart grids, concludes in its report "Smart Grids in Denmark" [6] that a smart grid is the most effective strategy to develop the electrical system and prepare it to meet future challenges. In this context, it is important to change the distribution networks, since distribution networks are the connection point between new agents with the grid [12]. The Ecuadorian electrical sector is currently pursuing new concepts like distributed generation such as micro- and peak hydraulic, photovoltaic, wind and biomass, and new loads such as electric vehicles and induction cookers. Encouraging customers for generating their energy is the key aspect considered within the new policies to convert Galapagos into an energetically self-sustaining archipelago.

14.2 Applied Methodology

After aforementioned, a compressive model of the entire grid is required, and this model must consider dynamics on load, the electromechanical-magnetic coupling between elements and thermal restrictions. Then, the low voltage modeling requires some steps as collect information of grid elements, its connection, and location, end users active and reactive power hourly profiles. Once this necessary information is gathered and validated, some scenarios are built, and in these scenarios different grid configurations are compared.

14.2.1 Modeling of LV Electric Network

In order to model various elements of the system such as lines, conductors, transformers [13], loads [14], and unconventional elements, e.g., photovoltaic panels, smart buildings, induction cookers, electric vehicles, and domestic wind turbines or photovoltaic arrays are presented on [10]. All these elements were modeled following the methodology described in [15]. Also, reference [16] presents evidence for considering the mutual effects between lines in unbalanced networks. Hence, it is essential to model all the lines considering the neutral wire. The MATLAB/Simulink environment has been used to model all the mentioned elements and for performing the simulations due to the ability to integrate real-time features when is incorporated RT-LAB.

14.2.1.1 Transformer

The Galapagos power system is mainly three-phase at the medium voltage level. Then, pole-mounted medium voltage/low voltage (MV/LV) transformers distribute electric power to end users. On the LV side of the transformer, a single-phase three-wire supply provides electricity at 120 and 240 V levels, as shown in Fig. 14.2.

The Galapagos network is equivalent to the North American networks topologically [17], and the transformer model parameters are detailed in [15].

Fig. 14.2 Equivalent circuit for center-tapped transformer

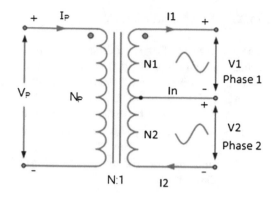

14.2.1.2 Photovoltaic Generation

Photovoltaic sources (PV) could be modeled from different approaches; for instance, in [18] the PV model, consider the single-diode five-parameter model. In [19], the block "three-phase dynamic load" fed with a negative profile is used. For this study, this last approach was selected since the need to evaluate the photovoltaic impact on the network using several photovoltaic curves from real measurements.

April is the month with the highest electric consumption in Galapagos; therefore, the active power measures of April of the existing PV power plant of 1500 kW installed on Santa Cruz Island were used to create 40 individual PV curves for the end users. The existing curves were transformed to per unit system after that; the desired rated power multiplied these. Figure 14.3 shows 40 real photovoltaic profiles.

According to [20], the energy resources in Santa Cruz Island are enough to satisfy the end users average consumption, and these resources even could generate 25% additional power. In order to achieve those mentioned above, it is necessary to install 2150 Wp at the roof of client's household. Normally, 2150 Wp are reached by means of a photovoltaic array composed of some panels.

Fig. 14.3 PV profiles

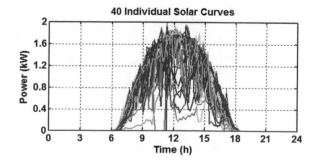

14.2.1.3 Electrical Motorbikes

The government of Ecuador desires to replace the conventional vehicles (fuel oil) by ecological and more efficient ones. Hence, a policy has been implemented to promote the massive change toward electric vehicles. A model of electrical motorbike taken from [20], with a rated power of 1.000 W, will be used. This profile is created upon a Monte Carlo Simulation. The maximum power is 630 W. See Fig. 14.4.

This average profile will be combined with a coincidence factor, which is a function of client's number [21]. The coincidence curve is defined in [20]. It is worth to mention that the electrical motorbikes, when charging, are connected to 240 V (phase-to-phase).

14.2.1.4 Loads

There are different types of clients in Galapagos, and most of them are residential; a typical residential client has various appliances such as a refrigerator, television, washing machines, microwaves, and lighting. The Galapagos utility makes regular measuring campaigns. Therefore, there are available real load curves concern both active and reactive power with 10 min samples [22]. Most are connected to 120 V (phase-to-neutral), and then, a random repartition of the 40 loads between phase 1 and 2 has been made (Figs. 14.5 and 14.6).

The approach that considers using real curves is much better than one which considers an average curve [21, 23].

14.2.1.5 Induction Cookers

In Ecuador, especially in Galapagos, several policies foster the change of conventional stoves by induction ones. The induction cooker's rated power is typically in the range 2500–7000 W. The voltage level is 240 V. Real measurements are used to feed the induction cooker model. The next figure shows 40 induction cooker profiles (Fig. 14.7).

Fig. 14.4 Electrical motorbike average curve

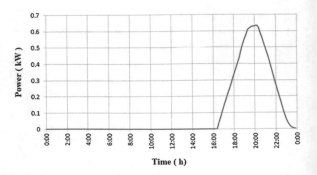

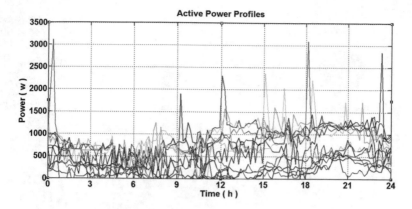

Fig. 14.5 Active power curves measured at end users

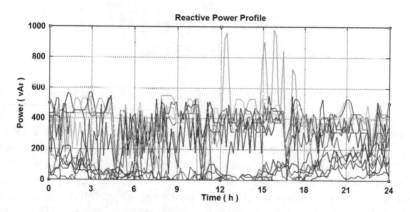

Fig. 14.6 Reactive power curves measured at end users

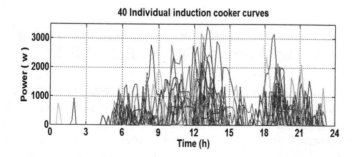

Fig. 14.7 Induction cooker profiles

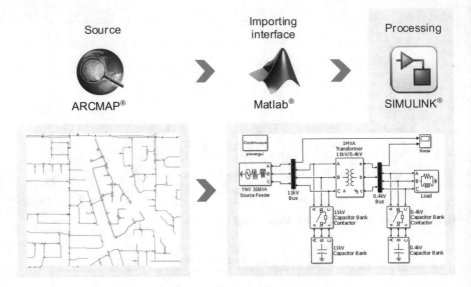

Fig. 14.8 Interface between ArcMap and MATLAB/Simulink

14.2.2 Interface Geographic Information Systems/Simulink

In order to limit the overall simulation time of the different scenarios, a software interface between the geographic information systems (GIS) (termed ArcMap) and MATLAB/Simulink has been developed. The logic consists into build the topology of the network inside the GIS. Afterward, using MATLAB code, the Simulink model is created automatically with all the connections between the elements. It is worth noting that, also, the technical information requested by Simulink is calculated in the script. See Fig. 14.8.

14.3 Impact Assessment of New Services in the Galapagos Low Voltage Network with Intelligent DSM

The advent of new services such as an electric motorbike, induction cooker, and distributed generator must be evaluated to measure their impact on the power network. The impact of the new services must be managed suitably, otherwise, this could bring worst results, for instance, loss of regular life of transformer due to overload [24, 25], higher frequency of outages, loss of reliability, and so on. Thus, realistic simulations are necessary to build a common understanding of the functions of an SG [26] in the particular case of Galapagos. These simulations also should seek to identify the impact on the grid of a high penetration of distributed generation like photovoltaic, which may cause bidirectional flows [27], and EV that usually represents a significant load.

14.3.1 Case Study

14.3.1.1 Description of the Study

Santa Cruz Island has three MV feeders at 13.8 kV, which are composed of three- and single-phase sections. To evaluate the impact of new services at the low voltage level, a MV/LV transformer with its network should be selected for the study. Hence, the information within the Geographical Information Systems has to be taken into account. First of all, the MV feeder 1 is chosen due to its residential characteristics. After that, statistical analysis has been performed on its LV sub-networks. Through the GIS's tools, a summary about the transformers with their number of clients has been made. Transformers with only one client were not considered. Once performed the analysis, 36 MV/LV transformers remain, the minimum number of customers per transformer is 2 and the maximum 40, the mean is approximately 17, and the standard deviation is 11.19 customers per transformer. Table 14.1 shows the main characteristics of the selected LV network, which is shown in Fig. 14.9.

14.3.2 Scenarios

To assess the impact of the new loads on the LV grid, some scenarios are defined. For each scenario, the voltages, power, and currents are analyzed. The nomenclature is PQ = load fed with active and reactive power curves, IC = induction cooker, EM = electrical motorbike, G = growing population and therefore of the load.

14.3.2.1 Scenario PQ

This scenario represents the current situation (reference case) of the selected LV network. Only the active and reactive load curves are considered.

Table 14.1 Description of the modeled LV network

Substation	Transformer	Power (KVA)	Customers	Nodes
Santa Cruz	TR1	50	40	15

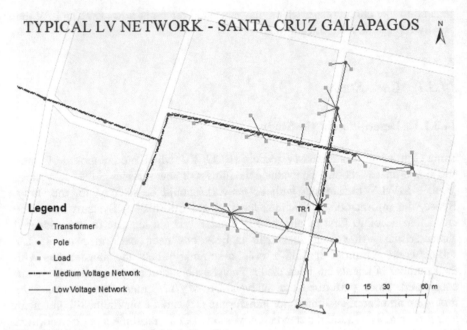

Fig. 14.9 LV network modeled into Simulink

14.3.2.2 Scenario PQ + IC

An induction cooker is implemented for each residential client, to assess the impact of IC on the grid variables.

14.3.2.3 Scenario PQ + IC + EM

An electrical motorbike for each client is added to the previous scenario.

14.3.2.4 Scenario PQ + IC + EM + G

This scenario considers an annually growing due to growing population rate leading to an 7.85% increase of the load [28]. This value is higher than growth population rate in the other provinces, due to the population has grown haphazardly mainly for the tourist activities. This growth has forced to increase the generation capacity and to deploy energy efficiency programs.

14.3.2.5   Scenario PQ + IC + EM + G + PV

In this scenario, assuming that the transformer will suffer overloads and take advantage of the existence of solar resources, an array of photovoltaic panels [20] is connected to each client house to 240 V.

14.3.2.6   Scenario Smart DSM

On the base of the scenario 5, a DSM program, which considers 15% of the controllable load. In this case, the smart function of Fig. 14.15 is considered for modifying the 40 load curves of the model.

14.3.3   Results

In the first scenario, the maximum power occurs at 20:48 and its value is 36,94 kVA. Regarding the drop voltage standard (limits ±5% of the nominal voltage), all the nodes are inside the margin. The maximum current through the transforms is 147,09 A, and the minimum voltage is 233,40 V (phase-to-phase). As expected, a current flowing through the neutral wire exists, and the maximum value is 27,06 A. This confirms that the network is unbalanced (Fig. 14.10).

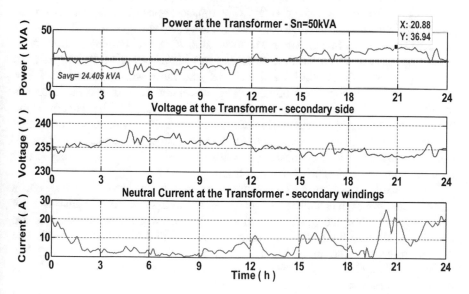

Fig. 14.10   Scenario PQ

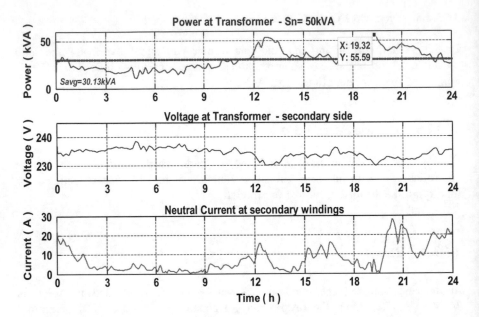

Fig. 14.11 Scenario PQ + IC

The results of the second scenario show an increase of 5,73 kVA in the average power. Now, the profile has two peaks during the day, the first one at 12:20 and the second one at 19:18. The peak in the night is higher and reaches a value of 56,05 kVA. The lowest voltage is 229,8 V (phase-to-phase). This scenario highlights overloads during the night and noon. However, voltages are still inside the standard limits (Fig. 14.11). The maximum value of neutral current is 28,15 A.

The scenario 3 shows an increase in the maximum power of 10,76 kVA, the second peak at night is 66,81 kVA at 19:18, and the lowest voltage is 228,2 V (phase-to-phase) during a few minutes. The maximum value of neutral current is 30,09 A, which indicates that the imbalance is still equivalent to the scenario 1 and 2. However, the maximum current in the transformer is 279,94 A, almost twice that for scenario 1. This scenario highlights voltages outside the standard limits and a significant overload in the transformer with a 134% relative load during a few minutes; Fig. 14.12 visualized this condition.

The scenario 4 also considers a population growth. As expected, the power curve of the transformer has undergone an increase, and its average value is now 33,77 kVA whereas the peak is 71,06 kVA. As in the scenario 3, voltages are outside the allowed limits; the minimum voltage is 227,2 V (phase-to-phase). The maximum neutral current is 35,15 A, and the maximum current through the transformer is 298.18 A. The relative load of the transformer reaches 142,12% as maximum during at least 30 min (Fig. 14.13).

The scenario 5 assesses the insertion of photovoltaic sources in the end user's facilities. Fig. 14.3 depicted that the solar resources are available since 6:00–18:00

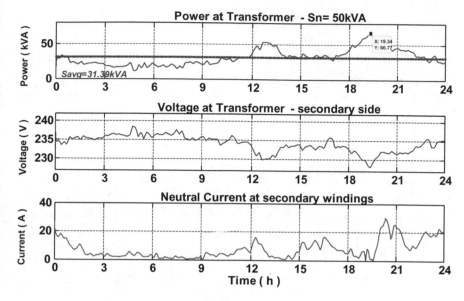

Fig. 14.12 Scenario PQ + IC + EM

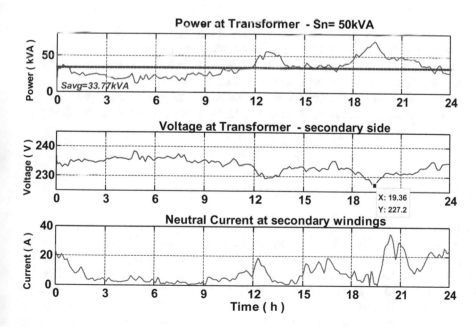

Fig. 14.13 Scenario PQ + IC + EM + G

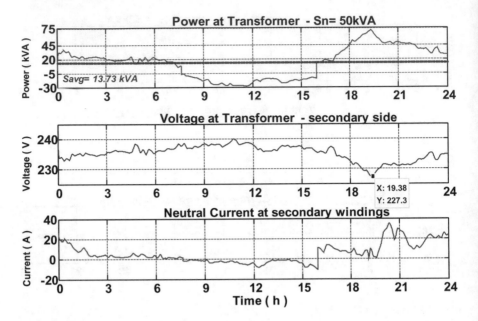

Fig. 14.14 PV scenario, where it is possible to see the reverse flow

approximately. The installed PV panels are enough (regarding produced power) to reverse the flow between 07:36 and 15:54. The average power in the transformer decreases considerably to 13,75 kVA. Despite the photovoltaic sources installation, we still have the power peak during the night for obvious reasons, and consequently, the voltage still crosses the lower limit during this peak. The maximum current flowing in the reverse sense is 122,44 A; the imbalance is practically the same as in the scenario 4 (Fig. 14.14).

14.3.4   Intelligent DSM for Galapagos Islands

In this section, an intelligent method that combines demand response (DR) and time of use (TOU), (DR + TOU), is presented. Nowadays, considering all the policies proposed by the government, the regulatory and control agency—ARCONEL—has updated the tariff schedule to include a new tariff for EV (up to 10kw) [29]. The primary initiative is based on fostering the EV deployment ought their efficiency and contribution to the environment as well as, the power grid applications such as (i) voltage control, (ii) exchange reactive power for autonomous voltage support without communicating with the distribution system operator, (iii) influencing the available active power for primary transportation function [30]. Unscheduled high penetration of EVs surely has adverse effects on power system performance when EVs used widely. As a result, there is an exigent need to predict the EVs' consumption. In [31], an approach to optimally operating microgrids in a context with renewable energy and EVs is addressed, and the approach takes into account a stochastic price of gas and electricity.

Assuming that 15% of the loads are controllable [32] and that every household possesses at least one controllable device which participates in the load management process, a smart function is created to represent the effect of applying TOU [33]. The 50% rebound effect defined in [34] is annulated using this function.

As an essential premise, this kind of DSM keeps the consumption almost equal, since exits only a load shifting [35]. This is presented in Eq. (14.1).

$$\int_0^{24} RpbTOU \approx \int_0^{24} RpwTOU \qquad (14.1)$$

where

RpbTOU is residential profile before TOU and RpwTOU is the residential profile with TOU.

The smart strategy has the same energy during the whole day with the difference that now, it will be considered an algorithm for the appliance controller device in each household and the shifted energy will be consumed during the whole day. Thus, the smart strategy will consider only actions over the burst loads and will use a modern DSM program (TOU + DR), thus no more rebound. See Fig. 14.15. In [36], an original energy management which uses a Virtual power plant is presented, exists only a demand resource considered; resistive water heaters, which are modeled like thermostatically controlled loads, however, the rebound effect is neglected.

As depicted, the consumption of the A region is shifted to the B region. The limit of the region B is 7:00 because after this time, the price is not the lower. Using the profiles affected by the smart function, another simulation is performed (Fig. 14.16).

As it was expected, until the scenario 4, the average power at the transformer grows. However, in the scenario 5, due to the installation of PV panels, this average

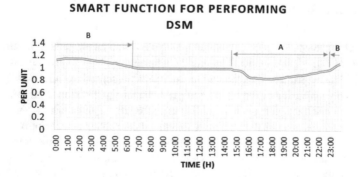

Fig. 14.15 Smart function for performing DSM

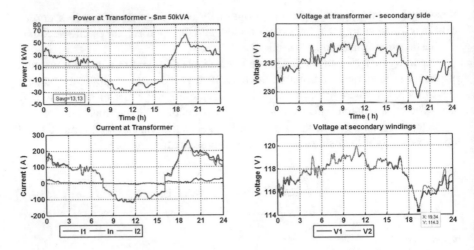

Fig. 14.16 Results after applying intelligent DSM

decreases by almost 60%, but we still have the same peak during the night. Thus, the scenario 6 is an intelligent DSM program, which can remove the peak created by the rebound and shift the energy during the whole day, keeping the average power at the transformer. The main results are presented in Table 14.2.

Regarding voltage levels, the scenario 6 appears to be the better, because it respects the regulation all the time, and reduction of the peak is around 10% compared with scenario 5. Also, regarding power overload of the transformer, it is a good solution because the value is "only" 128%. Indeed, according to [37], an acceptable value during two hours is 133%. In extreme cases, a distribution transformer could withstand a 150% overload for one hour [38, 39].

14.4 Application of Battery Energy Storage System to Decrease Relative Loads in Galapagos Distribution Transformers

The research presented in [33] considers Demand Size Management methods such as (i) time of use—TOU—and (ii) demand response—DR—to reduce overloads in the MV/LV transformers. The several simulations presented in [33, 35, 40] showed on one hand that during some hours the photovoltaic generation is higher than the load. On the contrary, the peak in Galapagos occurs in the night where the photovoltaic resources are not well enough to satisfy the power consumption; means of DSM methods reduce the peak during this time. However, it is quite clear that the photovoltaic surplus in the morning could be stored and used later in the peak hour through of battery energy storage systems (BESS) due to the ability to perform peak shaving and store energy for nighttime use. The storage elements are (i) batteries,

Table 14.2 Scenarios results

Scenario	Name	V1			V2			Power (kVA)			Over load
		Max	Min	Avg	Max	Min	Avg	Max	Min	Avg	Max
1	PQ	119,94	116,40	117,59	120,34	116,69	117,71	36,96	0,35	24,40	0,74
2	PQ + IC	119,94	114,84	117,08	120,34	114,88	117,21	56,05	0,35	30,13	1,12
3	PQ + IC + EM	119,94	114,06	116,99	120,34	113,99	117,13	66,81	0,35	31,39	1,34
4	PQ + Growing	119,94	113,58	116,71	120,34	113,48	116,87	71,06	0,35	33,77	1,42
5	PQ + PV	120,01	113,58	117,54	120,34	113,48	117,67	71,06	−28,38	13,73	1,42
6	DSM smart	120,01	114,37	117,59	120,34	114,31	117,72	64,07	−28,38	13,13	1,28

(ii) ultra-capacitors, (iii) compressed air, (iv) flywheels, (v) hydrogen fuel cell, and (vi) pumped hydro-storage have different applications [41].

The design and implementation of an advanced battery management system—BMS—is conducted in [42], with continuous monitoring of voltage and current of each battery. In addition, using DC–DC boost converter, a constant output voltage level is maintained. The research presented in [43] assesses the storage aging of different types of batteries such as Li-ion and lead–acid for stand-alone systems, to have a reliable lifetime estimation. A smart grid environment with a high penetration level of household's storage batteries is presented in [44], where the utilities influence family's consumption using electricity price structure, while the homes aim to reduce their bills using coordinate charging and discharging process. BEES are a promising technology for enabling the transition from traditional networks toward smart grids [45]. In [46], a proposal for a distributed cooperative control strategy for coordinating the energy storage system is presented, and the primary objective is to guarantee the supply–demand balance and minimize the total power loss associated with charging/discharging inefficiency. Whereas in [47] it is stated that for improving the reliability of the isolated microgrid, installing storage units is necessary. BESS can store energy during off-peak times and release the stored energy upon request (arbitrage). Reference [48] addresses a centralized multi-objective optimization algorithm to coordinate multi-microgrids optimally. The optimal operating point is a trade-off between minimization of operating cost, voltage profile deviation, and feeder congestion.

This section deals with battery optimal sizing determination in low voltage level to reduce the relative load in distribution transformers, and this objective represents an innovative research, since the current state of the art does not deal with the relative load target. A energy cost function minimization in presence of TOU tariff is applied considering the presence of photovoltaic generation. A case study in the low voltage network of Galapagos is analyzed; an innovative controller for respecting the different constraints is built in the MATLAB/Simulink environment. Finally, a suggested tariff for encouraging the battery installation within Galapagos Islands is derivate once realized all the simulations with the real information given by the utility.

14.4.1 Cost Functions Considering TOU Tariffs

A detailed model for determining the possibilities of storage in a context with TOU tariffs was presented in [49], although the given scenario does not consider the photovoltaic generation. In the case that the end user is not allowed to sell energy, and the total cost of the system C_T is given by

$$C_T(x) = C_0 + C_{en} \tag{14.2}$$

Co is the capital cost including purchase, installation, and maintenance (percentage of capital cost). The replacement cost is not included because the battery lifetime is not lower than the planning time, for our case (10 years) [50]. The energy cost C_{en} includes the expense of the load in the presence of photovoltaic generation $C_{load\ ppv}$, the expense of charging the battery C_{ch}, and the savings originated for discharging the battery C_{dch}. According to [51], the photovoltaic generation cost is considered as free.

$$C_{en} = C_{load\_ppv} + C_{ch} - C_{dch} \tag{14.3}$$

The model presented in [49] is adapted to take into account the photovoltaic generation and three price levels.

$$C_{load\_ppv} = \sum_{n=1}^{N} \left\{ \frac{1}{(1+\beta)^{n-1}} \cdot \left[D(i) \cdot \sum_{i=1}^{Sn} \begin{pmatrix} E_{L\_off}(n,i) - E_{ppv\_off}(n,i)) * \mathrm{Pri}_{off}(n,i) + \\ (E_{L\_med}(n,i) - E_{ppv\_med}(n,i)) * \mathrm{Pri}_{med}(n,i) + \\ (E_{L\_peak}(n,i) - E_{ppv\_peak}(n,i)) * \mathrm{Pri}_{peak}(n,i)) \end{pmatrix} \right] \right\} \tag{14.4}$$

where N is the number of years of the studied period and β is the discount rate. D(i) is the typical day's number of the ith season. E_{Loff}, E_{Lmed}, and E_{Lpeak} are the total energy absorbed by the load during the off-peak, medium consumption, and on-peak, respectively. $E_{ppv\ off}$, $E_{ppv\ med}$, and $E_{ppv\ peak}$ are the total energy generated by the PV panels during either off-peak or medium consumption or on-peak periods. Pri_{off}, Pri_{med}, and Pri_{peak} are the energy prices of off-peak, medium consumption, and on-peak periods of the ith season of the nth year.

In [52], the expression for cost is presented as follows.

$$C_{en} = Co + \sum_{n=1}^{N} \frac{1}{(1+\beta)^{n-1}} \left(\sum_{i=1}^{Sn} D(i) \left[A + e(n,i) \left(\frac{\mathrm{Pri}_{off}(n{*}i)}{\eta_{ch}} - \eta_{dch} {*} \mathrm{Pri}(n{*}i) \right) \right] \right) \tag{14.5}$$

with

$$A = \begin{bmatrix} E_{L\_off}(n,i) - E_{ppv\_off}(n,i)) * \mathrm{Pri}_{off}(n,i) + \\ (E_{L\_med}(n,i) - E_{ppv\_med}(n,i)) * \mathrm{Pri}_{med}(n,i) + \\ (E_{L\_peak}(n,i) - E_{ppv\_peak}(n,i)) * \mathrm{Pri}_{peak}(n,i)) \end{bmatrix} \tag{14.6}$$

Considering that photovoltaic generation reduces the energy consumption from the electric grid, the minimum cost is reached when the charge/discharge energy is maximal. Figure 14.17 depicts the system in analysis; the study takes into account battery degradation, power balance, and charging/discharging number per day.

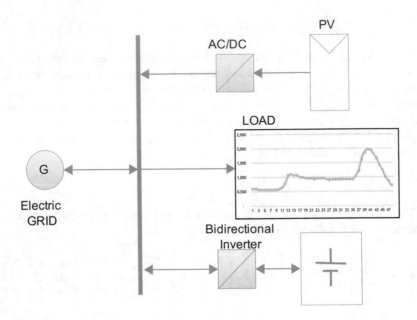

Fig. 14.17 System in study

Equation 14.7 represents the power balance in the system, where P_{EGRID} is the demanded power from the grid, P_{PV} is power supplied for PV, P_{BAT} is power supplied for battery systems, and P_{LOAD} is the required power for the load.

$$P_{EGRID}(t) = P_{LOAD}(t) + P_{PV}(t) + P_{BAT}(t))$$ (14.7)

14.4.2 Setting

The government regulation agency gives the TOU electricity tariffs, and ARCO-NEL updated the tariff schedule in order to include a new tariff for EV (up to 10 kw) [29]. Energy prices are $Pri_{off} = 0.05$ ¢/kWh from 22 PM to 8 AM, $Pri_{med} = 0.08$ ¢/kWh from 8 AM to 18 PM, and $Pri_{peak} = 0.10$ ¢/kWh from 18 to 22 PM. Figure 14.18 presents these prices.

Figure 14.19 depicts the load in each phase after DSM strategy, which assumed that 15% of the loads are controllable [32] and that every household possesses at least one controllable device which participates in the load management process, with a 50% rebound integrated into the simulation [34].

Between 7:23 and 16:06, a reverse flow is originated due to photovoltaic generation.

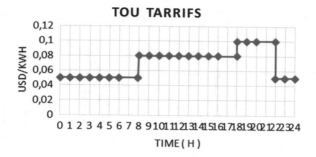

Fig. 14.18 TOU scheme

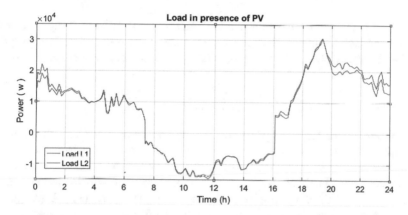

Fig. 14.19 Load in each phase in the presence of PV generation

14.4.3 Battery

First, the proper battery must be selected, nowadays; the two most developed types are (i) lead–acid and (ii) Li-ion. References [53, 54] present a detailed batteries review. The Li-ion battery possesses the greatest potential, a small size and low weight, higher energy density and efficiency near to 100%, lifespan (300 cycles at 80% of discharge), and self-discharge (1%/month). The main drawbacks are relational with investment cost around 731–1045 USD/kWh. The lead–acid (flooded type) battery has an efficiency $\eta = 72 - 78\%$, lifespan 200–300 cycles at 70% depth of discharge, self-discharge 2–5%/month, a relatively small cost 52–157 USD/kWh. A threshold value for assessing batteries in function of the prices, which takes into account the aging factor, is defined in [41] and presented in Eq. (14.8).

$$Kcrit \approx \frac{(max_{tarrif} - min_{tariff}) * \eta_B}{Z} \tag{14.8}$$

Fig. 14.20 Total cost for different batteries' sizes considering different TOU for batteries

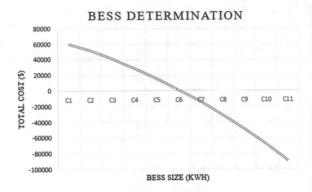

where Kcrit is the maximum value for the unit cost for the admitted capacity loss, Z is the aging factor, and η_b is the battery performance. According to the test proposed in [43], Z value is $3 * 10^{-4}$. Using the Galapagos data, in this case, Kcrit value is:

$$Kcrit \approx \frac{(10 - 5) * 10^{-5} * 0,9}{3 * 10^{-4}} \approx 0,15 \qquad (14.9)$$

Different values for K are presented in [55], for instance for lead–acid batteries k = 0,15 $/Wh, for Li-ion k = 1,33 $/Wh. According to the *Kcrit* and the *K* values for these two batteries types, the Li-ion battery would not be selected since the huge investment is not justified. The lead–acid battery was selected for the next simulations, although its *K* value is equal to the *Kcrit*. Thus, the next Assumption is defined: for the batteries considered such as a distributed energy resource: the consumers using batteries to reduce the peak will be beneficiaries of a TOU of 0,33USD/kWh. In addition, a fixed annual increment of 5% of the load is considered. In Galapagos, when the peak is reached, it is necessary to start thermal generators. Normally, the fee for those is about 0,40–0,45 USD/kWh. Thus, a new regulation of ARCONEL defines 0,57USD/kWh for a photovoltaic generation to encourage changing the technology [56] (Fig. 14.20).

The value that reaches a cost near to zero is C = 19,5 KWh.

14.4.4 Simulations

Figure 14.21 presents the load seen at transformer secondary side. The maximum demand is now 49,1 kVA due to the battery action. The section marked in red shows the reverse flow. As the reader can see, the constraint related to the assigned nominal power of the transformer Pn = 50 kVA is respected.

The maximum relative load is 98%; in [33], the minimum relative load achieved with Smart Techniques is 128%. Hence, a considerable reduction has been reached. Figure 14.22 presents the battery power; the controller starts the charging process when the photovoltaic generation is higher than the load until the production is equal to the load. After that, the discharge process begins. The discharging energy is consumed in less time (6,91 h) than the time used for charging.

Figure 14.23 presents the state of charge (SOC). The lower and upper levels are respected, as well as the constraint related to the number of cycles per day. Also, the maximum allowed depth of discharge is respected as well.

Figure 14.24 is presented to illustrate the behavior of the whole system better. In the point A, it is possible to see the beginning of the charging, and the point B is the end charging point. The battery and the electric grid feed the load at point C (peak), respecting the power balance. The point D is the end discharging point.

Fig. 14.21 Power at transformer (Phase 1 + Phase 2)

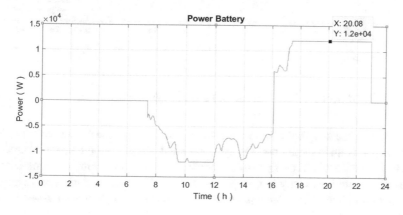

Fig. 14.22 Power battery connected at 240 V

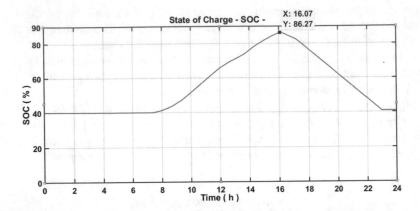

Fig. 14.23 SOC, one cycle per day

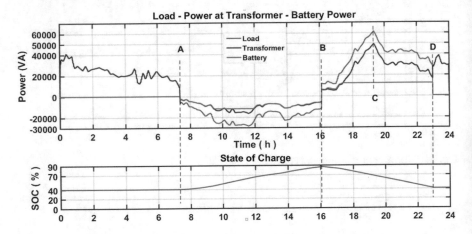

Fig. 14.24 Whole system behavior

14.5 Analysis in Networks on a Large Scale

This section deals with the simulation of the whole distribution network of Santa Cruz, and it is an essential point to get a detailed vision of the changes that would suffer the entire distribution system, considering the firm aim of implementing smart grid facilities shortly in Galapagos Islands. Thus, an analysis considering all the clients and the three feeders must be accomplished. Santa Cruz network is composed as follows:

1. There feeders (Medium voltage).
2. 465 MV/LV transformers.

3. 6625 customers: 5364 residential, 1118 commercial, 130 industrial, and 13 lighting streets.

The next figure depicts Galapagos Islands and the distribution network within Santa Cruz.

The electrical sector has a powerful interface to create the electrical model within CYMDIST [57] using the information coming from GIS and a database with electrical parameters. The next figure shows the whole network modeled into CYMDIST (Figs. 14.25 and 14.26).

Within CYMDIST, the low and medium voltage network is modeled. Hence, the impact in distribution transformers and on the whole system (Substation) is feasible. Figure 14.27 depicts the base scenario in Galapagos without any new charges; it means only active and reactive curves are considered. The maximum value in the substation is 6431 kVA.

Once defined the baseline scenario, it is possible to include in each client new services such as (i) EM, (ii) induction cooker, (iii) distributed generation, and (iv) TOU strategies in a massive way.

Figure 14.28 depicts the inclusion of an average profile of IC (coincidence factor considered) as well as the integration of EM and PV in a context with a TOU scheme. The next table illustrates the percentage of each element modeled.

With the values above mentioned in Table 14.3, a new simulation was performed, as the reader can see due to the DSM strategy the consumption is shifted between 22:00 and 4:00. The photovoltaic generation reduces the peak during noon. The maximum value in the substation is 7977 kVA (Fig. 14.29).

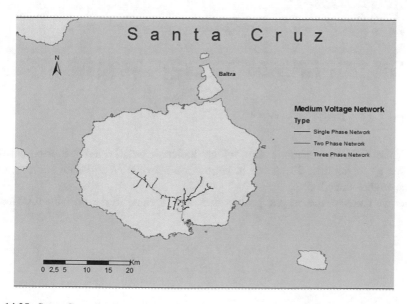

Fig. 14.25 Santa Cruz distribution network

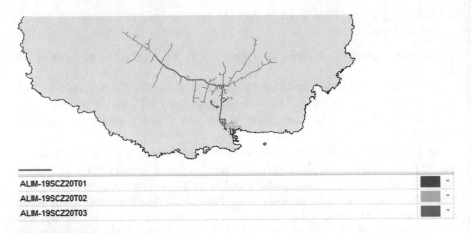

ALIM-19SCZ20T01	
ALIM-19SCZ20T02	
ALIM-19SCZ20T03	

Fig. 14.26 CYMDIST model

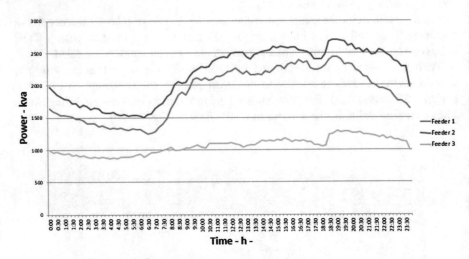

Fig. 14.27 Load profiles in the substation

Figure 14.30 is a thematic map, which depicts in detail the distribution network colored by loadability. The zones in yellow are between 90 and 95% and the zones in red higher than 105%.

As we can see, the biggest problems will be located in the populated center.

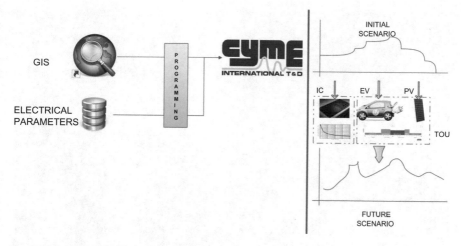

Fig. 14.28 Methodology applied on a large scale

Table 14.3 Values for assessing integration of new services

Item	Value (%)
Year	9
IC	70
EM	20
EM TOU	40
TOU residential	60
DGs resid	30

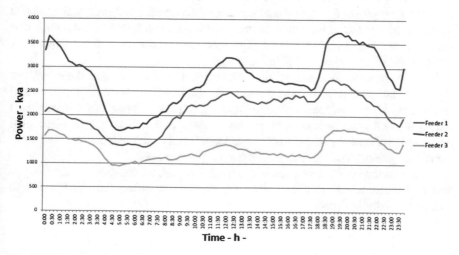

Fig. 14.29 New load profiles later 9 years

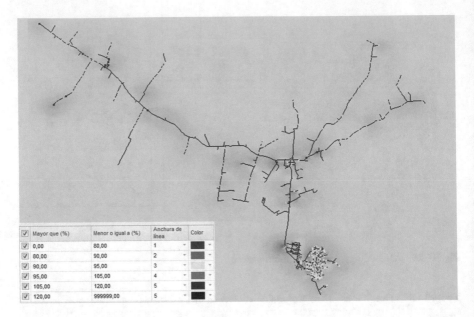

☑ Mayor que (%)	Menor o igual a (%)	Anchura de línea	Color
☑ 0,00	80,00	1	■
☑ 80,00	90,00	2	■
☑ 90,00	95,00	3	□
☑ 95,00	105,00	4	■
☑ 105,00	120,00	5	■
☑ 120,00	999999,00	5	■

Fig. 14.30 Loadability map

14.6 Conclusions

In the chapter, an example of a complex network is presented, and the Galapagos Archipelago electric network must operate overcoming some technical, environmental, and social issues.

In order to assess the impact of new services on the grid, a simulation environment is proposed, for that the elements connected to this network were modeled in detail. All specific information (location, connections, technical data, etc.) was extracted from geographic information systems; finally to make calculations, Simulink/MATLAB and CYMDIST are used.

Five scenarios were proposed; the first one is the reference case; the other four represent the integration of new services in the baseline scenario. Voltages and demand peak are analyzed, and from these results, it is clear that the proposed intelligent DSM works fine even in the worst scenario. The use of distributed generation reduces the daily load peak significantly and reduces the fuel consumption (primary combustible in electricity generation).

To incorporate storage systems in the modeled grid, a techno-economic analysis was performed, and results show that it is feasible to include battery systems to reduce the daily load peak.

References

1. Ecuador_Travel. *Galapagos Islands EcoTourism* (2016). http://www.ecuador.us/travel/galapagos_islands/info/galapagos_islands_ecotourism/
2. Google_Earth (ed.), *Galapagos's Satellite View* (2016). https://www.google.com/earth/
3. INEC, *Ecuador en Cifras—Población y Demografía* (2015)
4. AYESA, *EcoSmart Galapagos, A sustainable city*, Quito, Ecuador (2014)
5. M.H. Amini, B. Nabi, M.R. Haghifam, Load management using multi-agent systems in smart distribution network, in *2013 IEEE Power & Energy Society General Meeting*, 2013, pp. 1–5
6. F. Kamyab, M. Amini, S. Sheykhha, M. Hasanpour, M.M. Jalali, Demand response program in smart grid using supply function bidding mechanism. IEEE Trans. Smart Grid **7**, 1277–1284 (2016)
7. L. Mokgonyana, J. Zhang, L. Zhang, X. Xia, Coordinated two-stage volt/var management in distribution networks. Electr. Power Syst. Res. **141**, 157–164, 12 (2016)
8. P. Chittur Ramaswamy, J. Tant, J.R. Pillai, G. Deconinck, Novel methodology for optimal reconfiguration of distribution networks with distributed energy resources. Electr. Power Syst. Res. **127**, 165–176, 10 (2015)
9. United_State_Department_of_Energy, *What is the Smart Grid* (2016). https://www.smartgrid.gov/the_smart_grid/
10. R. Medina, Microrredes basadas en Electrónica de Potencia: Características, Operación y Estabilidad, in *Ingenius*, 2014, pp. 24–31
11. K.G. Boroojeni, M.H. Amini, S.S. Iyengar, Overview of the security and privacy issues in smart grids, in *Smart Grids: Security and Privacy Issues* (Springer International Publishing, Cham, 2017), pp. 1–16
12. D.X. Morales, R.D. Medina, Y. Besanger, Proposal and requirements for a real-time hybrid simulator of the distribution network, in *IEEE 2015 CHILEAN Conference on Electrical, Electronics Engineering, Information and Communication Technologies (CHILECON)*, 2015, pp. 591–596
13. R.D. Medina, J.P. Lata, D.P. Chacón, D.X. Morales, J.P. Bermeo, A.E. Medina, Health index assessment for power transformers with thermal upgraded paper up to 230 kV using fuzzy inference systems, in *Proceedings of 2011 49th International Universities' Power Engineering Conference (UPEC)*, 2016, pp. 1–8
14. M. Kezunovic, Teaching the smart grid fundamentals using modeling, simulation, and hands-on laboratory experiments, in *Power and Energy Society General Meeting, 2010 IEEE*, 2010, pp. 1–6
15. D. Morales, R. Medina, Y. Besanger, Proposal and requirements for a real-time hybrid simulator of the distribution network, *Chilecon-2015*, 2015
16. J. Sifuentes, Modélisation de réseaux de distribution dans un simulateur temps-réel pour des applications "Smart Grids", STAGE PFE, Institut Polytechnique de Grenoble, 2013
17. A. Dubey, S. Santoso, Electric vehicle charging on residential distribution systems: impacts and mitigations. Access, IEEE **3**, 1871–1893 (2015)
18. D. Sera, R. Teodorescu, P. Rodriguez, PV panel model based on datasheet values, in *IEEE International Symposium on Industrial Electronics, ISIE 2007*, 2007, pp. 2392–2396
19. A. Mercier, Pilotage de la production décentralisée et des charges non conventionnelles dans le contexte Smart Grid et simulation hybride temps réel. Energie électrique, PHD, G2ELab—Laboratoire de Génie Electrique de Grenoble, Université Grenoble Alpes, hal.archives-ouvertes.fr., 2015
20. Universidad Politécnica de Valencia, ANÁLISIS PARA LA IMPLEMENTACIÓN DE REDES INTELIGENTES EN ECUADOR, Insituto de Ingeniería Energética, Universidad Politécnica de Valencia, 2015
21. W.H. Lee, *Power Distribution Planning Reference Book*, vol. 2 (Marcel Dekker, Inc., 2004)
22. D. Morales, R. Medina, Real-time hybrid simulator of the distribution network for smart grid applications. SICEL 2015 **8**, 4, 20th Nov 2015

23. R. Medina, *Plan de gestión del consumo residencial en la Empresa Eléctrica Regional Centro Sur C.A* (Ing, Facultad de Ingeniería, Universidad Politécnica Salesiana, Cuenca, 2009)
24. R.D. Medina, A.A. Romero, E.E. Mombello, G. Ratta, Comparative study of two thermal aging estimating methods for power transformers. IEEE Latin Am. Trans. **13**, 3287–3293 (2015)
25. K. Bustamante, W. Borja, B. Miranda, L. Zhunio, R. Medina, Power transformers risk index assessment in the ecuadorian context, in *IEEE International Conference on Automatica (ICA-ACCA)*, 2016, pp. 1–6
26. G. Feng, L. Herrera, R. Murawski, E. Inoa, W. Chih-Lun, P. Beauchamp et al., Comprehensive real-time simulation of the smart grid. Ind. Appl. IEEE Trans. **49**, 899–908 (2013)
27. P. Kotsampopoulos, V. Kleftakis, G. Messinis, N. Hatziargyriou, Design, development and operation of a PHIL environment for Distributed Energy Resources, in *IECON 2012—38th Annual Conference on IEEE Industrial Electronics Society*, 2012, pp. 4765–4770
28. ARCONEL, *Estadisticas del Sector Electrico—Demanda Anual*, 2015
29. *ESQUEMA TARIFARIO PARA LA INTRODUCCIÓN DE VEHÍCULOS ELÉCTRICOS EN EL ECUADOR*, ARCONEL, 2015
30. K. Knezović, M. Marinelli, Phase-wise enhanced voltage support from electric vehicles in a Danish low-voltage distribution grid. Electr. Power Syst. Res. **140**, 274–283, 11 (2016)
31. S.M.M.H.N.S. Heydari, H. Mirsaeedi, A. Fereidunian, A.R. Kian, Optimally operating microgrids in the presence of electric vehicles and renewable energy resources, in *2015 Smart Grid Conference (SGC)*, 2015, pp. 66–72
32. Universidad Politécnica de Valencia, ANÁLISIS PARA LA IMPLEMENTACIÓN DE REDES INTELIGENTES EN ECUADOR: Definición Cuantitativa, Instituto de Ingenieria Eléctrica, Quito, Ecuador, 2015
33. D.X. Morales, Y. Besanger, S. Sami, C.A. Bel, Assessment of the impact of intelligent DSM methods in the galapagos Islands towards a smart grid. Electr. Power Syst. Res. **146**, 308–320 (2017)
34. Greenlys_Project, *Un premier retour sur expérience*, 2015. http://greenlys.fr/
35. D.X. Morales, Y. Besanger, M. Toledo, R.D. Medina, Impact study of new loads and time of use schedule in the low voltage Network, in *IEEE PES Innovative Smart Grid Technologies, Europe* Ljubljana-Slovenia, 2016
36. L.A.D. Espinosa, M. Almassalkhi, P. Hines, S. Heydari, J. Frolik, Towards a macromodel for packetized energy management of resistive water heaters, in *2017 51st Annual Conference on Information Sciences and Systems (CISS)*, 2017, pp. 1–6
37. FACILITIES ENGINEERING BRANCH DENVER OFFICE, permissible loading of Oil-Immersed transformers and regulators, in *Facilities Instructions, Standards, and Techniques,* vol. 1–5, ed. Denver, Colorado, 2000, p. 28
38. IEEE recommended practice for performing temperature rise tests on Oil-Immersed power transformers at loads beyond nameplate ratings, in *IEEE Std C57.119-2001,* p. 0_1, 2002
39. IEEE, IEEE guide for loading Mineral-Oil-Immersed transformers and step-voltage regulators, in *IEEE Std C57.91-2011 (Revision of IEEE Std C57.91-1995)*, 2012, pp. 1–123
40. D.X. Morales, Y. Besanger, C. Alvarez, R.D. Medina, Impact assessment of new services in the galapagos low voltage network, in *IEEE PES T&D Transmission and Distribution Latin America*, Morelia-Mexico, 2016
41. Y. Ru, J. Kleissl, S. Martinez, Storage Size Determination for Grid-Connected Photovoltaic Systems. IEEE Trans. Sustain. Energ. **4**, 68–81 (2013)
42. A.T. Elsayed, C.R. Lashway, O.A. Mohammed, Advanced Battery Management and Diagnostic System for Smart Grid Infrastructure. IEEE Trans. Smart Grid **7**, 897–905 (2016)
43. E. Lemaire-Potteau, F. Mattera, A. Delaille, and P. Malbranche, Assessment of storage ageing in different types of PV systems: technical and economical aspects, in *23rd European Photovoltaic Solar Energy Conference and Exhibition*, Valencia, Spain, 2008, pp. 2765–2769
44. C.O. Adika, L. Wang, Non-cooperative decentralized charging of homogeneous households' batteries in a smart grid. IEEE Trans. Smart Grid **5**, 1855–1863 (2014)

45. C. Brivio, S. Mandelli, M. Merlo, Battery energy storage system for primary control reserve and energy arbitrage. Sustain. Energy Grids Netw. **6**, 152–165, 6 (2016)
46. Y. Xu, W. Zhang, G. Hug, S. Kar, Z. Li, Cooperative control of distributed energy storage systems in a microgrid. IEEE Trans. Smart Grid **6**, 238–248 (2015)
47. B. Falahati, A. Kargarian, Y. Fu, Timeframe capacity factor reliability model for isolated microgrids with renewable energy resources, in *2012 IEEE Power and Energy Society General Meeting*, 2012, pp. 1–8
48. A. Kargarian, M. Rahmani, Multi-microgrid energy systems operation incorporating distribution-interline power flow controller. Electr. Power Syst. Res. **129**, 208–216, 01 Dec 2015
49. G. Carpinelli, F. Mottola, D. Proto, Probabilistic sizing of battery energy storage when time-of-use pricing is applied. Electr. Power Syst. Res. 73–83 (2016)
50. PowerThru, *Lead Acid Battery working—Lifetime Study*, 29 Dec 2016. http://www.power-thru.com/documents/The%20Truth%20About%20Batteries%20-%20POWERTHRU%20White%20Paper.pdf
51. Y. Riffonneau, S. Bacha, F. Barruel, S. Ploix, Optimal power flow management for grid connected PV systems with batteries. IEEE Trans. Sustain. Energ. **2**, 309–320 (2011)
52. D.X. Morales, Développement de la gestion optimale de l'énergie électrique dans les îles Galápagos vers les Réseaux Intelligents (PhD, UNIVERSITE GRENOBLE ALPES, France, 2017)
53. K.C. Divya, J. Østergaard, Battery energy storage technology for power systems—an overview. Electr. Power Syst. Res. pp. 897–905 (2009)
54. S. Faias, J. Sousa, R. Castro, Contribution of energy storage systems for power generation and demand balancing with increasing integration of renewable sources: application to the Portuguese power system, in *2007 European Conference on Power Electronics and Applications*, 2007, pp. 1–10
55. B. University, *BU 403: Charging Lead Acid*, 2016. http://batteryuniversity.com/learn/article/charging_the_lead_acid_battery
56. *PRECIOS DE LA ENERGÍA PRODUCIDA CON RECURSOS ENERGÉTICOS RENOVABLES NO CONVENCIONALES*, ARCONEL, 2006
57. CYME, *Análisis de Sistemas de Distribución*, 07 July 2017. http://www.cyme.com/es/software/cymdist/

Index

A

Adaptive channel allocation, 93
Adoption, 6, 77, 122, 124, 126–129, 131
Applications of networked control systems, 80
A Solar Test Bed, 20
Attack detection, 79, 85, 91, 92
Autonomous response, 160, 168, 169

B

Bidding strategy, 247
Bi-directional communication, 123, 235
Bio-inspired algorithm, 142, 144

C

Campus sustainability, 15, 16
Centralized and decentralized approach, 234, 235
Centralized approach, 161, 235
Centralized System, 169
Charging optimization, 6, 136, 140, 143, 144
Climate change, 13, 15, 17–19
Climate leadership, 17
College instructors, 5, 47
Communication, 5, 21–32, 34, 35, 37, 40, 42, 79–85, 89, 91, 92, 108, 124, 126, 137, 151, 153, 235, 253
Community detection, 98–102, 105, 107
Complex networks, 1, 2, 4, 97, 98, 198, 278
Compressed Sparse Row (CSR), 68
Computational intelligence, 6, 136, 139
Consumer behavior, 127
Control, 2, 3, 5, 22, 23, 29, 31–34, 39–41, 77–80, 82, 83, 86, 87, 91, 92, 124, 126, 135, 138, 139, 151, 201, 222, 234, 236
Coordinated response, 162, 164

D

Decentralized energy systems, 173
Demand response, 141, 151–153, 160, 166, 168–170, 233, 234, 264, 266
Demand side management, 138, 222, 253
Distributed approach, 4, 165–168
Distributed generation, 151, 155, 170, 234, 253, 259, 275, 278
Distributed intelligence, 34
Distributed power systems, 222
Distributed system, 164, 169
Dynamic clustering, 115
Dynamic network, 98, 99, 101, 102, 108, 109

E

Electric vehicle, PEV, 135, 136
Electric Vehicles (EV), 6, 22, 26, 36, 37, 121, 131, 135, 136, 138, 139, 141, 156, 170, 222, 235, 253, 254, 256
Energy expense, 154, 159, 161, 162, 165, 167, 168
Environmental concern, 233

F

Fog computing, 24, 29, 31

G

Galapagos islands, 252, 253, 268, 274, 275
Genetic algorithm, 39, 141, 154, 223, 225–227, 231
Geographic information systems, 258, 278
GPU, 67, 68, 70–75
Graph, 67–75, 98, 99, 101, 102, 104, 109, 111
Graph representation, 67, 68
Grid-to-vehicle, 22, 25, 36

© Springer International Publishing AG, part of Springer Nature 2018
M. H. Amini et al. (eds.), *Sustainable Interdependent Networks*,
Studies in Systems, Decision and Control 145,
https://doi.org/10.1007/978-3-319-74412-4

H
High performance, 67, 75
Home-wide management, 168
Household Energy Management (HEM), 234

I
IEEE 30-Bus Test System, 231
IEEE 9-bus and 30-bus test systems, 231
Intelligent communication, 25, 42
Interdependent networks, 1–5, 24, 35, 40, 42
IoE, 22
IoT, 22, 25–28, 30, 35, 41, 42

L
LiDAR, 174, 189–192, 209, 211–213
Load frequency controller, 86
Low voltage modeling, 254

M
Machine learning, 174, 193, 194, 196–198,
      205, 206, 208, 209, 211–213, 236
Massachusetts Institute of Technology, 14
Methodology review, 6
Multi-agents, 234, 247
Multi-GPU, 74

N
Networked Control Systems (NCS), 80
Non-sequential approach, 167–169
Non-sequential System, 167
Networks, 1–6, 15, 21–26, 28, 31, 33–35,
      40–42, 67, 82, 86, 99–105, 107–111,
      114, 115, 127, 137, 139–141, 180,
      196–198, 207, 222, 251, 253, 254, 259,
      268, 274

O
Occupational burnout, 5, 47, 48, 50, 51, 53, 55,
      59, 60
Optimization, 1–4, 22, 23, 27, 35, 40, 41, 79,
      124, 135, 136, 138, 139, 141, 142, 144,
      154, 162, 198, 200, 203, 204, 223, 268
Organizational commitment, 5, 47–49, 51, 53,
      54

P
Parallel, 30, 67, 69, 70, 75, 143, 222
Peak rebound, 160–163, 165, 168, 169

PHEV, 121–131, 138
Photovoltaic, 2, 139, 173, 191, 253–255, 259,
      262, 266
Photovoltaic generation, 243, 255, 266, 268,
      269, 272, 273, 275
Physical and empirical models, 175, 176, 210
Planning for Climate Resiliency, 18
Plug-in Electric Vehicle (PEV), 6, 135, 136,
      234
Power electronic devices, 2, 129, 222
Power purchase agreement, 17–19
Power systems, 4, 6, 24, 35, 36, 38–41, 78–80,
      86, 138, 213, 221–223, 253

Q
Q-learning, 236
Q-learning algorithm, 234, 236, 237, 239, 247

R
Real power losses, 222, 223, 225, 228, 230,
      231, 268
Real-world networks, 104, 110
Reinforcement learning, 236, 237, 247
Renewable energy generators, 20
Residential consumer, 151, 152, 158, 161, 164,
      168, 169
Residential demand, 151, 153, 169, 170

S
Sampling methods, 174, 187, 188, 190, 212,
      213
Scalable, 13, 21, 35, 42, 100, 115
Security, 4, 5, 24, 26, 28, 30, 31, 33–36, 41,
      77–79, 82, 85, 92, 122, 126, 128, 222,
      228
security, 228, 230, 231, 234
Seed-Centric approach, 100
Sequential approach, 168, 169
Sequential system, 165
SIMT, 67
Simulink, 254, 258, 260, 268, 278
Smart city, 1, 2, 4, 21–28, 30, 31, 33–37,
      39–42
Smart energy design, 25, 31, 35, 42
Smart grid, 4, 6, 24, 32, 35, 122–124, 143, 151,
      213, 221, 234, 253, 268, 274
Smart power grids, 2
Social network analysis, 97

Solar global horizontal models, 176
Solar mapping, 208
Solar tilted radiation models, 176, 177
Static VAR Compensator (SVC), 199,
    223–231
Sustainable development, 14
Sustainable interdependent networks, 2–4, 6
System-wide coordination, 152, 160, 161, 164,
    169, 170

T
Time of use, 31, 33, 264, 266

Transforming systems, 20
Transportation networks, 1, 4

V
Variable load, 240, 242, 244, 246
Vehicle-to-grid (V2G), 6, 22, 25, 36, 122–131,
    136
Vehicle-to-vehicle, 22, 25, 36
Voltage Profile Index, 224
Voltage profile index, 228, 229
Vulnerability assessment, 18

Printed in the United States
By Bookmasters